Lecture Notes in Chemistry

Edited by G. Berthier, M. J. S. Dewar, H. Fischer,
K. Fukui, H. Hartmann, H. H. Jaffé, J. Jortner,
W. Kutzelnigg, K. Ruedenberg, E. Scrocco, W. Zeil

7

Ion Cyclotron
Resonance Spectrometry

Edited by
Hermann Hartmann
Karl-Peter Wanczek

Springer-Verlag
Berlin Heidelberg New York 1978

Editors

Hermann Hartmann
Akademie der Wissenschaften
und der Literatur zu Mainz
Geschwister-Scholl-Straße 2
D-6500 Mainz

Karl-Peter Wanczek
Institut für Physikalische Chemie
der Universität Frankfurt
Robert-Mayer-Straße 11
D-6000 Frankfurt am Main 1

Library of Congress Cataloging in Publication Data

Hartmann, Hermann, 1914-
    Ioncyclotron resonance spectrometry.

    (Lecture notes in chemistry ; 7)
    Bibliography: p.
    Includes index.
    1. Ion cyclotron resonance spectrometry. I. Wan-
czek, Karl-Peter, 1940-   joint author. II. Title.
QD96.I54H37   543'.085   78-5862

ISBN 978-3-540-08760-1   ISBN 978-3-642-93085-0 (eBook)
DOI 10.1007/978-3-642-93085-0

softcover reprint of the hardcover 1st edition 1978

2152/3140-543210

Preface

In this volume, for the first time, a dozen of papers is collected dealing with almost all important aspects of ion cyclotron resonance spectrometry. The ICR technique was developed very rapidly in the last two decades. It seems to the editors that the method is now established well enough to dedicate a progress report to it.

This report is devided into three parts: The first articles present new developments in the theory of ICR spectrometry. They are followed by papers on recent developments of the experimental technique. About half of the volume is dedicated to applications of ICR spectrometry to reactivity, reaction mechanism and structure in the chemistry of thermal ions.

The editors are indebted to Mrs. A. Tin for typewriting most of the manuscript and to Mrs. E. Jacke for reproduction of Figures appearing in the volume.

Hermann Hartmann,
Akademie der Wissenschaften und der Literatur
Mainz

Karl-Peter Wanczek
Institut für physikalische Chemie
Universität, Frankfurt/Main

Contents

Line Shapes in ICR Spectra

by

A.H. Huizer and W.J. van der Hart
Department of Theoretical Organic
Chemistry, University of Leiden
P.O.Box 75, Leiden, The Netherlands

1. Introduction

Line shapes of ICR-signals as obtained in an idealized ICR cell have been discussed in several papers. Here we refer especially to the papers by Buttrill [1] and Comisarow [2]. Buttrill discussed line shapes for non-colliding ions by solving the equation of motion:

$$m \frac{d\vec{v}}{dt} = q\{\vec{E} + \vec{v} \ \vec{B}\}$$

whereas Comisarow took account of collisions by introducing a damping term $-c\vec{v}$ as proposed before by Beauchamp [3]. This results in the equation of motion for the averaged ion velocity:

$$m \frac{d\vec{v}}{dt} = q\{\vec{E} + \vec{v} \ \vec{B}\} \ -c\vec{v} \tag{1}$$

The derivation of line shapes from this equation, especially for reactive ions, is rather complicated and leads to expressions which are difficult to interpret.

Now, ICR is in many aspects comparable to magnetic resonance where line shape expressions are commonly derived in a much simpler way by application of linear response theory.

It is seen in eq. 1 that in ICR the velocity of the ions depends linearly on the electric field and it has been shown in a quantum mechanical treatment [4a, 4b] that eq. 1 also holds for the corresponding Heisenberg operators.

In the following sections we will show that even in rather complicated cases line shapes, peak heights and peak areas can be derived in a straight-forward and simple way by application of linear response theory.

2. The ICR Signal

Before deriving expressions for line shapes it is essential to deter-mine which property of the ions is measured in ICR. In previous papers this was generally assumed to be the energy absorbed from the perturbing r.f. field. However, ICR signals can also be obtained in cases where there is no energy absorption at all. Fourier-Transform ICR[5] is a well-known example of such an experiment. We therefore consider the generation of ICR signals in more detail (c.f. ref. 4b).

In the usual continuous wave ICR spectrometer ions are accelerated by a r.f. field

$$E_x(t) = E_0 \cos \omega t$$

(the coordinate axes are defined in Fig. 1). As a result the average ion velocity in the x direction becomes:

$$<v_x(t)> = v_1 \cos \omega t + v_2 \sin \omega t \tag{2}$$

The forced motion of the ions induces a signal current in the upper and lower plates of the analyzer proportional to $<v_x(t)>$:

$$i_s(t) = i_1 \cos \omega t + i_2 \sin \omega t \tag{3}$$

The analyzer plates are part of a resonant LC-circuit (Fig. 1) which is fed from a current source $i_0 \cos \omega t$ to generate the required r.f. electric potential $V_0 \cos \omega t$. The total voltage in the LC-circuit then becomes (see Fig. 1):

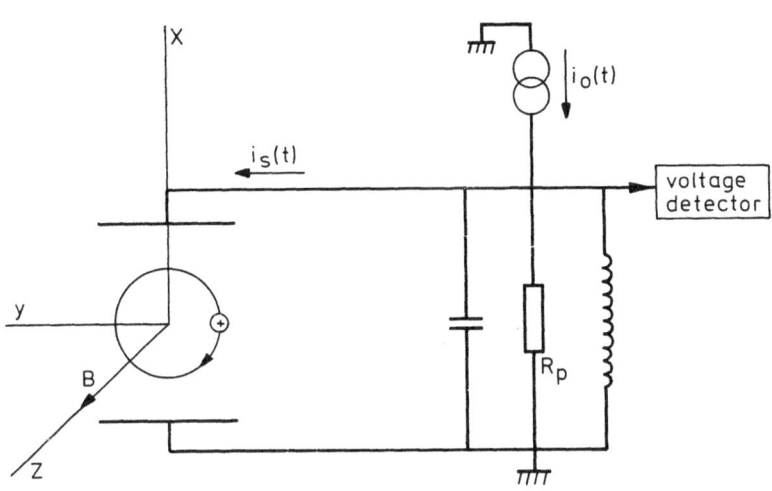

Fig. 1 Basic diagram of an ICR spectrometer.

$$V(T) = V_0\cos \omega t - i_1 R_p \cos \omega t - i_2 R_p \sin \omega t$$

in which R_p is the equivalent parallel resistance of the LC-circuit. In commonly used ICR spectrometers the amplitude of this voltage:

$$V_d = \{(V_0 - i_1 R_p)^2 + (i_2 R_p)^2\}^{1/2}$$

is detected. Because V_0 is much larger than the signal voltage, V_d can be approximated by:

$$V_d = V_0 - i_1 R_p \tag{4}$$

It follows that only the i_1 component of the signal current is measured.

In an electric circuit the average rate of energy absorption is given by the product of voltage and current averaged over a cycle of the r.f. field so the rate of energy absorption of the ions becomes:

$$A = \overline{V(t)\ i_s(t)}^C = \tfrac{1}{2} V_0\ i_1 \tag{5a}$$

The magnitude of A can also be found as the cycle-averaged product of the electric force e $E_x(t)$ and the average ion velocity in the x-direction $v_x(t)$ multiplied by the number of ions N_0 present in the analyzer:

$$A = N_0 q\ \overline{E_x(t)\ <v_x(t)>}^C$$

which, by substitution of eq. 2, yields

$$A = \tfrac{1}{2} N_0 q\ E_0\ v_1 \tag{5b}$$

Since E_0 is proportional to V_0, it follows from eqs. 5a and 5b that

$$i_1 = \alpha N_0 q v_1 \tag{6}$$

in which α is a proportionality constant determined by the cell geometry. It follows from eq. 5a that the rate of power absorption, discussed in previous papers, is proportional to the measured quantity i_1. In a general discussion of ICR phenomena it is, however, much more convenient to consider $i_s(t)$ as the signal. First of all, as mentioned above, ICR can be observed in cases when the perturbing oscillator is switched off as in FT-ICR. In the second place, it follows from eq. 1 that there is a linear relation between the ion velocity and the accelerating electric field strength. According to eq. 6 this means that the ICR signal intensity is proportional to E_0 whereas the rate of energy absorption is proportional to E_0^2. In our experience experimental signal intensities are indeed proportional to the oscillator voltage.

Moreover, it follows from the linear relation between signal current
and accelerating field strength that ICR line shapes can be derived in a
general way by application of linear response theory.

3. Application of Linear Response Theory

3.1. Outline of the Method

As shown in section 2, the signal current $i_s(t)$ at time t depends
on the average ion velocity $<v_x(t)>$ at time t. The ion velocity itself is
obtained by the action of the accelerating electric field E(t) at earlier
times. In a discussion of signal shapes it is then important to know how
the signal current at time t depends on the value of E(t') at time t'. We
therefore consider the electric field to consist of a series of rectangular pulses of height E(t)' and duration $\Delta t'$. Because of the linearity condition the contribution $\Delta i_s(t)$ due to one of these pulses should be proportional to both E(t') and $\Delta t'$ and its variation in time should depend
on (t-t') only. It follows that:

$$\Delta i_s(t) = R(t-t') \ E(t') \ \Delta t'$$

In this expression R(t-t'), which is called the response function, describes the behaviour of the signal current after an acceleration of the
ions caused by a σ-function electric field pulse at time t'.

The total current response at time t is just the sum of the individual
response $\Delta i_s(t)$ and can be found by integration over all previous times:

$$i_s(t) = \int_{-\infty}^{t} R(t-t') \ E(t') \ dt'$$

In the normal c.w. ICR spectrometer $E(t) = E_0 \cos \omega t$ which yields:

$$i_s(t) = \int_{-\infty}^{t} R(t-t') E_0 \cos \omega t' \, dt'$$

By the change of variables, $\tau = t-t'$, this expression is transformed into:

$$i_s(t) = E_0 \cos \omega t \int_0^{\infty} R(\tau) \cos \omega \tau \, d\tau + E_0 \sin \omega t \int_0^{\infty} R(\tau) \sin \omega \tau \, d\tau$$

in accordance with eq. 3. From the discussion in section 2 it then follows that ICR line shapes can be obtained from the first term

$$g(\omega) = E_0 \int_0^{\infty} R(\tau) \cos \omega \tau \, d\tau \qquad (7)$$

Consequently, in a derivation of line shapes one only needs to know the response function. Because of the relation between the original current and the average ion velocity in the x direction, the response function can be obtained from the behaviour of the ion velocity $<v_x(t)>$ after a very short electric field pulse. Application of a pulse E_p in the x direction of duration Δt at time $t = 0$ produces an average ion velocity

$$<v_x(0)> = \frac{q}{m} E_p \Delta t$$

Due to the cyclotron motion $<v_x(t)>$ will oscillate with the cyclotron frequency, so, provided the ion motion is not interrupted, one obtains:

$$<v_x(t)> = \frac{q}{m} E_p \Delta t \cos \omega_0 t$$

with ω_0 being the cyclotron frequency qB/m.

In practice, the amplitude of the average ion velocity will decrease in time. Ions may change their velocity in direction and magnitude due to elastic or inelastic collisions.

Therefore we may write

$$\langle v_x(t) \rangle = \langle v'_x(t) \rangle \cos \omega_0 t \qquad (8)$$

with $\qquad \langle v'_x(0) \rangle = (q/m)E_p \Delta t,$

to take account of the decay of the average velocity.

Besides, the signal current after an electric field pulse to the analyzer section decreases in time because ions escape observation due to drift out of the analyzer, unimolecular or collision induced reactions and losses by collisions with the walls of the cell. All these effects can be taken together by use of a decay function $R'(t)$ defined by

$$R(t) = R'(t) \cos \omega_0 t$$

with $\qquad R'(0) = \dfrac{q^2}{m} N_0.$

where we have omitted the proportionality constant α because it only depends on the cell geometry.

Substitution in eq. 7 gives:

$$g(\omega) = E_0 \int_0^\infty R'(t) \cos \omega_0 t \cos \omega t \, dt \qquad (9a)$$

Close to resonance $(\omega_0 - \omega) \ll \omega$ and because $R'(t)$ varies slowly compared to $\cos(\omega + \omega_0)t$ we may neglect terms in $(\omega + \omega_0)t$ which gives

$$g(\omega) = \frac{E_0}{2} \int_0^\infty R'(t) \cos(\omega - \omega_0) t \, dt \qquad (9b)$$

From eq. 9b it is easy to see that the peak height is given by:

$$I_{max} = g(\omega_0) = \frac{E_0}{2} \int_0^\infty R'(t) dt \qquad (10)$$

whereas the peak area is derived as

$$I = E_0 \int_0^\infty g(\omega) d\omega = E_0 \frac{\pi}{2} R'(0) = \frac{E_0 \pi q^2 N_0}{2m} \qquad (11)$$

This latter result is especially interesting because it clearly shows that peak areas are independent of the collision frequency as shown before in a somewhat less direct way

In the following sections we discuss the decay processes in more detail. First of all we consider non-colliding, non-reactive ions in a normal drift cell. Then we will introduce collisions and finally we will take account of chemical reactions.

3.2. Non-reactive Ions in a Drift Cell in the Low Pressure Limit

In the case of non-reactive ions the density of each type of ion is constant over the length of the cell. Due to the drift of ions through the cell the current response after an electric field pulse to the analyzer region will decrease in time and becomes zero when all the ions, present at the time of the pulse, are drifted out of the analyzer. It follows that R'(t) has the form:

$$R'(t) = \frac{q^2 N_0}{m} (1 - t/t_d) \quad 0 \leq t \leq t_d \tag{12}$$

$$R'(t) = 0 \qquad\qquad t > t_d$$

with t_d being the analyzer drift time (Fig.2).

The number of ions N_0 present in the analyzer region is given by

$$N_0 = P t_d \tag{13}$$

with P being the rate of ion production in the source.

By substitution of 12 and 13 in 9b one obtains

$$g(\omega) = \frac{P t_d q^2 E_0}{2m} \int_0^{t_d} (1 - t/t_d) \cos(\omega - \omega_0) t\, dt$$

and consequently,

$$g(\omega) = \frac{q^2 E_0 P}{2m} \frac{\{1 - \cos(\omega - \omega_0) t_d\}}{(\omega - \omega_0)^2} \tag{14}$$

which is identical to the line-shape expression for non-reactive ions given by Buttrill multiplied by a factor $2/E_0$.

The peak height can be found by setting $\omega = \omega_0$ in eq. 14 or from eq. 10:

$$I_{max} = \frac{N q^2 E_0}{2m} \int_0^{t_d} \{1 - t/t_d\}\, dt$$

$$= \frac{q^2 E_0 P t_d^2}{4m} \tag{15}$$

In the derivation of eqs. 14 and 15 it is assumed that the r.f. field is ideal, i.e. it is constant over the analyzer region and does not extend into the neighbouring sections of the cell. In practice, there will of course be a gradual change of the r.f. field at the beginning and end of the analyzer. As a result the response function also loses its sharp edges at $t = 0$ and $t = t_d$. (See Fig. 2).

For the line shape function this implies that the oscillations in the wings of the peak will be smoothed out.

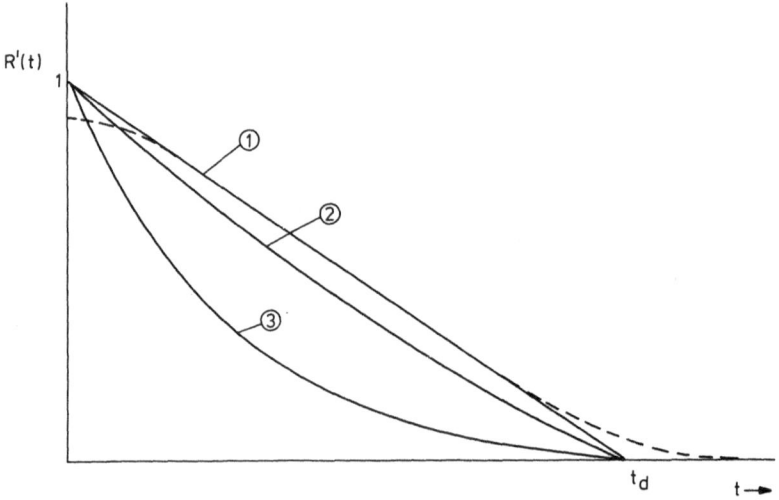

Fig. 2. The response amplitude $R'(t)$ under various conditions
    curve 1: primary ions, no reactions, no collisions
    curve 2: secondary ions, $t_s = t_d/2$, $k_1 = 1/t_d$, $c = 0$, $k_j = 0$
    curve 3: primary ions, $c = 1/t_d$, $k_1 = 2/t_d$
    All curves scaled to $R'(0) = 1$

3.3. Introduction of Collisions

As shown in eq. 1 collisions are usually introduced by a damping term $-cv$ in the equation of motion. Due to this term average ion velocities will decrease exponentially with time. So eq. 12 must be replaced by (Fig. 2):

$$R'(t) = (q^2 N_0/m) \exp(-ct) \{1 - t/t_d\}, 0 \leq t \leq t_d \qquad (16)$$

$$R'(t) = 0 \qquad t > t_d$$

and the line shape becomes

$$g(\omega) = \frac{Pt_d q^2 E_0}{2m} \int_0^{t_d} \exp(-ct) \{1 - t/t_d\} \cos(\omega - \omega_0)t \, dt$$

which yields:

$$g(\omega) = \frac{q^2 E_0 P}{2m\{c^2 + (\omega - \omega_0)^2\}} \left[ct_d + \frac{1}{\{c^2 + (\omega - \omega_0)^2\}} \right.$$

$$\times \left\{ -2c(\omega - \omega_0) \exp(-ct_d) \sin(\omega - \omega_0)t_d + \{(\omega - \omega_0)^2 - c^2\} \right\} \Bigg]$$

$$\times \{1 - \exp(-ct_d) \cos(\omega - \omega_0)t_d\} \tag{17}$$

which is the line-shape expression given by Comisarow, again multiplied by a factor $2/E_0$.

The peak height is found either from eq. 10 or from eq. 17 to be:

$$I_{max} = \frac{q^2 E_0 P}{2mc^2} \{ct_d - 1 + \exp(-ct_d)\} \tag{18}$$

3.4. Chemical Reactions

The effect of chemical reactions on the response function and consequently on the line shape is twofold.

First of all a reaction is an additional path by which the response function decays. It follows that instead of the damping coefficient c used in the previous section we now have

$$c' = c + k' + nk''$$

in which k' represents the unimolecular dissociation of ions and k'' is the total rate constant for reactions with neutral gas molecules.

A second consequence of reactions is a non-uniform distribution of ions over the length of the cell. Therefore the decay of the response function due to the drift of ions out of the cell is no longer linear with time as in eqs. 12 and 16. Instead one has to calculate which part of the ions which were present at the time of the electric field pulse would still be present at time t if they were non-reactive.

Before doing so, we first determine the ion distribution by transforming the time dependence of ion concentrations into a y dependence with y being the position of an ion in the cell measured from the electron beam (Fig. 1). So, for primary ions the number of ions per unit cell length becomes:

$$N_1(y) = \frac{P}{v_d} \exp(-k_1 y / v_d) \tag{19}$$

with v_d being the drift velocity and k_1 the total reaction rate constant of the primary ion,

and for secondary ion j:

$$N_j(y) = \frac{P}{v_d} \frac{k_{1j}}{k_1 - k_j} \{\exp(-k_j y / v_d) - \exp(-k_1 y / v_d)\} \tag{20}$$

in which k_{1j} is the rate constant for the formation of secondary ion j and k_j is its total reaction rate constant.

For a general case, the response function now becomes:

$$R'(t) = (q^2/m) \exp(-c't) \int_{y_1}^{y_2 - v_d t} N(y) dy \qquad (21)$$

where y_1 and y_2 are the positions of begin and end of the analyzer respectively. These boundaries can also be defined in terms of t_d and t_s with t_s being the time the ions need to drift through the source region.

$$y_1 = v_d t_s$$

$$y_2 = v_d (t_s + t_d)$$

The response function for primary ions becomes

$$R'(t) = (q^2 P/m) \exp(-k_1 t_s - ct - k_1 t) \frac{1 - \exp(k_1 t - k_1 t_d)}{k_1} \qquad (22)$$

and for secondary ions

$$R'(t) = (q^2 P/m) \frac{k_{1j}}{k_1 - k_j} \exp(-ct)$$

$$x \left\{ \exp(-k_j t_s - k_j t) \times \frac{1 - \exp(k_j t - k_j t_d)}{k_j} - \exp(-k_1 t_s - k_j t) \right.$$

$$\left. x \frac{1 - \exp(k_1 t - k_1 t_d)}{k_1} \right\} \qquad (23)$$

The corresponding line shape expressions can be derived by substitution of eqs. 22 and 23 in eq. 9b. The resulting formulae are rather complicated (c.f. ref. 2) so we will restrict ourselves to the peak heights and areas. Application of eqs. 10 and 11 gives for primary ions

$$I_{max} = (q^2 P/2m) \; E_0 \; \exp(-k_1 t_s)$$

$$x \left\{ \frac{1 - \exp(-k_1 t_d - ct_d)}{k_1(c + k_1)} - \exp(-k_1 t_d) \; \frac{1 - \exp(-ct_d)}{k_1 c} \right\} \quad (24)$$

and

$$I = (q^2 P/2m) \; \pi E_0 \; \exp(-k_1 t_s) \; \frac{1 - \exp(-k_1 t_d)}{k_1} \qquad (25)$$

For secondary ions

$$I_{max} = (p^2 P/2m) \; E_0 \; \frac{k_{1j}}{k_1 - k_j}$$

$$x \left[\exp(-k_j t_s) \left\{ \frac{1 - \exp(-k_j t_d - ct_d)}{k_j(c + k_j)} - \exp(-k_j t_d) \; \frac{1 - \exp(-ct_d)}{k_j c} \right\} \right.$$

$$- \exp(-k_1 ts) \left\{ \frac{1 - \exp(-k_j t_d - ct_d)}{k_1(c + k_j)} - \exp(-k_1 t_d) \right.$$

$$\left. \left. x \; \frac{1 - \exp(-ct_d - k_j t_d + k_1 t_d)}{k_1(c + k_j - k_1)} \right\} \right] \qquad (26)$$

and

$$I = (q^2 P/2m) \; \pi E_0 \; \frac{k_{1j}}{k_1 - k_j}$$

$$x \left\{ \exp(-k_j t_s) \; \frac{1 - \exp(-k_j t_d)}{k_j} - \exp(-k_1 t_s) \; \frac{1 - \exp(-k_1 t_d)}{k_1} \right\} (27)$$

The derivation for primary and secondary ions given here can easily be extended to more general reaction schemes.

If we can derive from the kinetical parameters a steady state distribution of the ions along the cell, application of eq. 21 gives the response amplitude $R'(t)$.

The derivation of the line shape properties is then straightforward.

References

1. S.E. Buttrill, J. Chem. Phys. 50, 4125 (1969)

2. M.B. Comisarow, J. Chem. Phys. 55, 205 (1971)

3. J.L. Beauchamp, J. Chem. Phys. 46, 1231 (1967)

4.a. P.W. Atkins and M.J. Clugston, Mol. Phys. 31, 603, (1976)

   b. A.H. Huizer and W.J. van der Hart, Mol. Phys. 33, 897 (1977)

5.   M.B. Comisarow and A.G. Marshall, J. Chem. Phys. 64, 110 (1976)

6.   W.J. van der Hart, Chem. Phys. Lett. 23, 93 (1973)

Quantum-Mechanical Description of Collision-Dominated

Ion Cyclotron Resonance

Hermann Hartmann, Kyu-Myung Chung
Institut für Physikalische Chemie der Johann-Wolfgang-Goethe
Universität, Robert-Mayer-Str. 11, 6000 Frankfurt (Main)

Introduction

Most ICR experiments are carried out at pressures of the order of 10^{-6} Torr and involve a swarm of ions moving through a dilute neutral molecule gas under the influence of the crossed homogeneous electric and magnetic fields which are superimposed by the trapping field which has the direction of the magnetic field and the alternating electric field between the drift plates.

At low pressures a few ion-neutral collisions occur during the observation time and the ions absorb the energy from the rf electrical field with frequency ω at a sharp frequency $\omega = \omega_c$; a finite absorption linewidth arises from instrument effects, which are mainly due to the electrical inhomogeneity in the ICR cells arising from the application of a trapping potential having the direction of the magnetic field. The electric and magnetic inhomogeneities affect the resolubility of the ICR spectrometers. The inhomogeneity effects on the resolubility have been investigated elsewhere by Hartmann et al. (to be published) [1]. At high pressures the ions undergo many collisions with neutrals during their transit through ICR cells, but with a collision frequency much less than ω they absorb energy over a range of frequencies near ω_c; the finite linewidth then arises primarily from collisions. The half width measurements of collision-broadened ICR spectral lines would offer possibilities for obtaining detailed information about nonreactive ion-neutral potentials if we could calculate the collision cross section for some model potentials for interaction with a few adjustable parameters.

It is argued [2] that a complete theory of collision-broadened ICR line shapes should start with the Boltzmann equation [3,4] for the ion distribution function, but most of the features of the experiment can be described by a phenomenological equation of Langevin [5] for an average ion, in which all the collisional effects appear in a single damping term characterized by ξ, the collision frequency for momentum transfer. One usually gets such a phenomenological equation of motion through two steps, i. e., multiplying the Boltzmann equation with an arbitrary function of ion velocity and integrating over the velocity space, one obtains the moment equation [6,7]. Applying the quasi-steady approximation to the moment equation, which is justified for the collision-dominated ICR experiments, and taking ion velocity as the factor function itself one obtains finally the phenomenological Langevin equation of motion of an average ion.

Kihara [8] has applied for the first time the moment method to the mathematical theory of electrical charges in gases. He has introduced an arbitrary function as deviation from the Maxwell distribution function and has replaced this arbitrary function by the eigenfunctions of the linear collision operator, which are defined in terms of Sonine polynomials. Recently Viehland and Mason have applied a completely arbitrary function of ion velocity to form the moment equations and they have chosen as basis functions the (Burnett-like) spherical polar functions, which are closely related with the Sonine polynomials chosen by Kihara except the quantity

T_i as a temperature parameter instead of the temperature T by Kihara.

Viehland et al.[2] have chosen again in their kinetic theory of ion cyclotron resonance collision broadening the same basis functions, which were
originally taken for the problem of gaseous ion mobility in electric fields
alone.

It is essential how one tries to solve this kind of phenomenological
moment equation. Beauchamp[9] was the first to attempt to treat the collision broadening of ICR lines for ion-neutral interactions of the form
(12-6-4) potential. In order to treat the Boltzmann equation he made two
assumptions, namely (1) the moment involving the collision operator is
equal to the product of moments, and (2) the ion distribution function can
be represented by an equilibrium function depending on an effective temperature that has to be calculated from the Boltzmann equation. Viehland et
al. have expanded the quantity which arises from the application of the
collision operator on the arbitrary function of ion velocity in terms of
the complete set of Burnett-like functions as usually done in kinetic theory.

The practical success of this method strongly depends upon the choice
of basis functions; if the test eigenfunctions happen to be similar to the
true eigenfunctions of the collision operator, convergence will usually
be rapid. The matrix elements of the collision operator, which are the
expansion coefficients of moments for collision operator multiplied by
an arbitrary function of ion velocity, can be expressed in terms of parameters T_i of dimension of temperature and the conventional collision integrals[3, 4], which depend upon the ion-neutral interaction. The evaluation of matrix elements, which is the most laborious part of this method,

is tedious and the calculation of generalized transport cross sections and the collision integrals cannot be carried out. Thus the matrix elements can only be reduced to linear combinations of irreducible collision integrals.

Recently Atkins and Clugston [10] have presented a quantum-mechanical approach based on a projection-operator formalism for the ion motion and the instantaneous power absorption in the ICR. A purely quantum-mechanical description of an ion motion in the ICR cells is not sufficiently adequate, because very high quantum numbers are involved due to $\hbar\omega_c$ <<kT and the Landau energy levels are lying closely together. Therefore, the system shows a nearly classical behaviour. In order to treat this nearly classically behaving system quantum-mechanically Hartmann and Chung [11] have applied a minimized wavepacket approach to the motion of an ion in the ICR cells. The wavepacket is constructed with the aid of the exact wavefunctions for this problem [12] and is minimized according to the uncertainty principle. This procedure corresponds to the problem of going from quantum theory to classical theory [13]. The quantum-mechanical pure states are quite different from the classical one. However, from the infinite superpositions of pure states one can construct the so-called coherent states in which the dispersion of various physical quantities takes its minimum value. Such states come closest to the classical states in such a way that one can actually follow the classical orbits of the ion motion in ICR cells, for example.

The purpose of this paper is to present a quantum theory for the collision broadening of ICR absorption lines, based on the minimized wavepacket approach. Without applying the Boltzmann transport-equation the phenomeno-

logical Langevin equation including the Lorentz force will be derived from
the Neumann average of velocity of ions with the aid of the statistical
density operator [14]. The latter is constructed using the minimized wavepacket solutions of the ICR problem. In this way we avoid a certain arbi-
trariness which entered in the conventional kinetic theory of ion cyclotron resonance collision broadening. Moreover, specifying the form of the
ion-neutral potentials with adjustable parameters for the non-reactive
collisions the collision rates can be expressed in a closed form. In this
way we can determine the dependence of collision rates on the various
factors rigorously.

Derivation of the Phenomenological Equation of Motion of an Ion in the ICR Cell

The ion moving under the influence of crossed electric and magnetic
fields superimposed by the trapping field parallel to the magnetic field
direction and capable of colliding with background neutrals absorbs the
energy from the rf field applied between the drift plates. The dominant
feature of this motion can be described neglecting the translational motion of the center of mass and disregarding the trapping field, which
causes the instrument effects. The Hamiltonian for an ion of reduced mass
μ is, if we use the symmetrical gauge for the vector potential $\vec{A} = \vec{B}/2 \times \vec{r}_\perp$

$$H = \frac{1}{2\mu} \left(\vec{p} - \frac{q}{c} \vec{A} \right)^2 - qEy + H_{im} + H_1(\tau) \tag{1}$$

where H_{im} means the ion-neutral interaction and $H_1(\tau) = - qE_1 \cos\omega\tau \, e^{n\tau}$

In order to derive the phenomenological Langevin equation of motion of an
average ion without using the Boltzmann equation we apply the minimized

wavepacket solution constructed with the exact wave functions for the Hamiltonian $H_0 = H - H_{im} - H_1(\tau)$ with the same symbols as in [11] except the drift velocity $v_D = c|\vec{E}|/|\vec{B}|$

$$\Psi_\perp(r_\perp,t) = \sqrt{\frac{b}{2\pi}(\frac{b}{2})^n \frac{1}{n!}} \, \exp\{(i/\hbar)\mu(\vec{V}_D,\vec{r}_\perp) + (iq/\hbar c) \, (\vec{A}(r_M),\vec{r}_\perp)\}$$

$$\cdot \, \exp\{-(b/4)(\vec{r}_\perp - \vec{r}_M)^2\} \cdot |\vec{r}_\perp - \vec{r}_M|^n \cdot \exp\{-in\varphi\} \tag{2}$$

$$\cdot \, \exp\{-(i/\hbar)[(n + \tfrac{1}{2})\,\omega_c - (qE/2)y_M + (\mu/2)v_D^2\,]t\}$$

As has been carried out in [11] the minimized wave packet solution turns out to be

$$\Psi_\perp = \sum_n P_n \, \Psi_\perp (\vec{r}, \vec{r}_M) \tag{3}$$

with the weightfunction

$$P_n = (n!e)^{-1/2} \, \exp\{in(\alpha + \pi/2)\} \, I_n(\varkappa, \sqrt{2/b}) \tag{4}$$

Linearizing the density operator $\rho = \sum_\nu |\nu\rangle W_\nu \langle\nu|$, which is constructed with the eigenvectors of the Hamiltonian $\bar{H}_0 = H_0 + H_{im}$, with regard to H_{im} we replace $|\nu\rangle$ by their zeroth and first order approximations:

$$\rho_0 = \sum_m |m\rangle W_m \langle m| \tag{5}$$

$$\rho_1 = \sum_m \{|m\rangle W_m \langle m|H_{im}(H_0 - E_m - i\hbar\zeta)^{-1} + h\,c\,\}$$

We denote the eigenstate vectors for $\bar{H}_0$ with Greek letters ν and that for H_0 with Roman letters m, which represent all of the characteristics for the wave packet state.

Applying the gauge invariant derivatives $\partial_j \equiv \nabla_j - (iq/\hbar c)\vec{A}_j$ $(j = x,y)$ the Neumann mean value can be represented with the aid of the density operators ρ_0 and ρ_1 for the Heisenberg velocity operator as

$$\langle v_j(t)\rangle = \sum_1 \langle 1|(\partial_j\rho_0)(t)|1\rangle \, (\hbar/i\mu) + \sum_1 \langle 1|(\partial_j\rho_1)(t)|1\rangle \, (\hbar/i\mu) \tag{6}$$

As we can deduce from the structure of the statistical density operator

(5), the first term in (6) does not concern the collisions and the second

term is responsible for the ion-neutral collisions. Therefore, we can make

use of the Heisenberg equations of motion as follows $(\partial_j \rho_0)$ (t) and

$(\partial_j \rho_1)$ (t) respectively

$$\frac{d}{dt}(\partial_j \rho_0)(t) = (1/i\hbar)[(\partial_j \rho_0) \ (t), \ H_0 + H_1 \ (\tau)] \tag{7}$$
$$\frac{d}{dt}(\partial_j \rho_1)(t) = (1/i\hbar) \ [(\partial_j \rho_1) \ (t), \ H_{im}]$$

This result agrees with the assumption that the matrix elements for odd

power of H_{im} vanish [11]. The time derivative of the Neumann mean value

is then employing of the relations (7) and the commutation relations

$$[\partial x, \partial_y^2 \ ] = -(2\mu^2/\hbar^2) \ \omega_c \ v_y$$
$$[\partial_y, \partial_x^2 \ ] = +(2\mu^2/\hbar^2) \ \omega_c \ v_x \tag{8}$$
$$[\partial_y, \ y \ ] = 1$$

It is given by

$$\frac{\partial}{\partial t} <v_j(t)> = (q/\mu)\{(E + E_1 \cos\omega\tau \cdot e^{\eta\tau}) \ \delta_{j,y} + [\vec{<v>}/c \times \vec{B}]_j\} \tag{9}$$

$$+ \ (1/i\hbar)\sum_l <1|[(v_j \rho_1) \ (t), \ H_{im}]|1>$$

The last term in (9) corresponds to the damping term in the Langevin equa-

tion and reduces to

$$(1/i\hbar)\sum_l <1|[(v_j \rho_1) \ (t), \ H_{im}]|1> \tag{10}$$

$$=-(1/n_i^+ V)(2\mu/\hbar)\sum_{l,m} <m|v_j \rho_0|m><m|H_{im}|1>|^2 W_{ml} \delta(E - E_{1,m})$$

because of the orthogonal relations of wave packet solutions and the re-

lations

$$\lim_{t \to \infty} \frac{e^{(i/\hbar)E_{1m}t}}{E_1 - E_{m \mp i\hbar\varepsilon}} = \pm 2\pi i\delta(E - E_{1m}) \qquad \cdot$$

We then obtain with the abbreviations $F(\tau) = (E + E_1 \cos\omega\tau e^{\eta\tau})\delta_{j,y}$

$$\langle vj\rangle_m = \langle m|v_j \rho_o|m\rangle \quad \text{and} \quad \xi_{m1} = (1/n_i^\dagger V)(2\pi/\hbar)|\langle m|H_{im}|1\rangle|^2 W_{m1}\delta(E - E_{1m})$$

the phenomenological equation of the motion of ions in ICR cells without making use of the Boltzmann transport equation and without any assumption with regard to the ion-molecule collisions:

$$\frac{\partial}{\partial t}\langle v_j\rangle = (q/\mu)\{F(\tau) + [\langle\vec{v}\rangle/c \times \vec{B}]_j\} - \sum_{m,1} \langle v_j\rangle_m \xi_{m,1} \tag{11}$$

This equation is formally the same equation of motion of an average ion in ICR cells as Beauchamp[9] has obtained factoring his collision term. Viehland et al.[2] have claimed that their collision operator J_j has the eigenfunction v and the eigenvalue ξ_j adapting the Maxwell model for the distribution of ion velocities. Namely, they have obtained with $\xi = \sum_j N_j \xi_j$ the decoupled equation of motion usually used as the starting point for conventional theories of ICR.

In contrast to the kinetic theory of Viehland et al. and of Beauchamp the collision frequency can be calculated for the appropriate ion-neutral potentials. A set of simpler coupled differential equations for the mean values $\langle v_j\rangle$ (11) can be decoupled through the analytical evaluation procedure, in which we utilize the orthogonality of the minimized wave packet and the relation between the classical rotational energy and that of quantized, namely

$$\mu\omega_c^2 R^2/2 = (1 + {}^1\!/_2)\hbar\omega_c$$

The details for this decoupling procedure will be given by the evaluation of ξm1 for the concrete form of ion-neutral potentials. This procedure gives the decoupled equations for the components of the mean value of ion velocity with which we construct a differential equation for the

$v_{\mp} = v_x \mp i v_y$, namely

$$\frac{\partial}{\partial t} <v_{\mp}> + (\hat{\xi} \mp i\omega_c) <v_{\mp}> = \mp (iq/\mu) F(\tau) \tag{12}$$

where the symbol $\hat{\xi}$ indicates the collision frequency after the decoupling procedure described above.

Solution of the Equation of Motion and Calculation of Power Absorption

The equation of motion (12) can be solved and has the solution of the form

$$<v_y> = (q E_y/\mu) \, Re \, (\frac{\hat{\xi} - i\omega_c}{\hat{\xi}^2 + \omega_c^2}) \, \{1 - exp[-(i\omega_c + \hat{\xi})t]\}$$

$$+ (qE_1/2\mu) \, Re \, \{(\frac{\hat{\xi} - i(\omega + \omega_c)}{\hat{\xi}^2 + (\omega + \omega_c)^2}) \, [1 - exp(-i(\omega + \omega_c) + \hat{\xi})t] e^{i\omega t} \}$$

$$+ (qE_1/2\mu) \, Re \, \{(\frac{\hat{\xi} - i\delta\omega}{\hat{\xi}^2 + \delta\omega^2}) \, [1 - exp(-i\delta\omega + \hat{\xi})t] e^{i\omega t} \} \tag{13}$$

Out of this solution we obtain for the mostly used ICR experimental conditions $\hat{\xi} \ll \omega$ and $\delta\omega = \omega - \omega_c \ll \omega + \omega_c$ for the signal currents in the measuring plates

$$j_y = \frac{n_i^+ q^2 E_1}{2\mu} \, Re \, \{(\frac{\hat{\xi} - i\delta\omega}{\hat{\xi}^2 + \delta\omega^2}) \, [1 - exp(-i\delta\omega + \hat{\xi})t] e^{i\omega t} \} \tag{14}$$

The instantaneous power absorption of a swarm of n_i^+ ions may be calculated from the relation

$$A(t,\omega) = E(t) j_y \tag{15}$$

and is given by

$$A(t,\omega) = \frac{\hat{\xi} q^2 n_i^+ E_1^2 cos^2\omega t}{2\mu(\hat{\xi}^2 + \delta\omega^2)} \, \{1 + e^{-\hat{\xi}t}[(\delta\omega/\hat{\xi})sin\delta\omega t - cos\delta\omega t]\}$$

$$- \frac{\hat{\xi} q^2 n_i^+ E_1^2 sin2\omega t}{4\mu(\hat{\xi}^2 + \delta\omega^2)} \, \{(\delta\omega/\hat{\xi}) - e^{-\hat{\xi}t}[(\delta\omega/t)cos\delta\omega t + sin\delta\omega t]\} \tag{16}$$

This expression contains two types of components - a high-frequency component given by a factor of $\sin 2\omega t$ or $\cos^2\omega t$, and a low-frequency one given
by $\sin\delta\omega t$ or $\cos\delta\omega t$. If we average the high-frequency component over the
cycloidal period $T = 2\pi/\omega_c$, we obtain for the power absorption from the
surviving term with $\cos^2\omega t$ replaced by its average value of 1/2

$$A(t,\omega) = \frac{\hat{\xi}\alpha^2 n_i^+ E_1^2}{4\mu(\hat{\xi}^2 + \delta\omega^2)} \quad \{1 + e^{-\hat{\xi}t} [(\delta\omega/\hat{\xi})\sin\delta\omega t - \cos\delta\omega t]\} \tag{17}$$

This function leads to the Buttrill[15] and Beauchamp[9] spectral functions for the low and high pressure limits respectively.

To complete the discussion of collision-dominated ICR we consider the
average excess kinetic energy of a swarm of ions undergoing power absorption at high pressures which is given by the energy gained from the alternating electric field minus the average energy lost in collisions. For
this purpose we form the phenomenological equation for the average of the
velocity squared again using the linearized statistical density operator
with respect to ion-neutral interaction H_{im} and applying the Heisenberg
equation of motion. A similar procedure as for the case of the average velocity gives the equation

$$\frac{\partial}{\partial t} <v^2> = \frac{2q}{\mu} <v_y> F(\tau) - \hat{\xi} <v^2> \tag{18}$$

Under steady state conditions the left side of equation (18) vanishes. The
first term on the right side of equation (18) is related to the power absorption defined by (15) such that converting the reduced mass to the ion
mass one gets

$$\hat{\xi} <v^2> = (2/m) (A(t,\omega)/n_i^+) \tag{19}$$

For further discussion we utilize the relations between the $\hat{\xi}$ and the collision rates in high temperature approximation $\hat{\xi} = (2\pi/\hbar)\sum_{m,i}|\langle m|H_{im}|1\rangle|^2\delta(E-E_{m1})$ in [11], which has been obtained converting the difference of distribution functions in (10) $W_{m1} = W_m - W_1$ into the difference of velocities or of velocities squared. This conversion is readily done through exchange of the role of eigenstate vectors in the last term. Thus we obtain

$$\hat{\xi} = \frac{\langle\Delta v\rangle}{\langle v\rangle}\,\bar{\xi},$$

$$\hat{\xi} = \frac{\langle\Delta v^2\rangle}{\langle v\rangle}\,\bar{\xi}. \tag{20}$$

For the elastic collision both collision frequencies coincide. So long as the energy exchange is taking place only between rf fields and ions and thereafter between ions and neutrals elastically, we can utilize the high temperature approximation expression as collision frequencies. For the inelastic collisions we should distinguish both frequencies, because the energy exchange between ions as well as internal energies of neutrals can be involved during the collision process.

Converting the energy in the center of mass frame to the laboratory system and making use of the relation between kinetic energy of an ion before the collision and the energy partition after the collision, we obtain

$$\frac{1}{2}m\,\langle v_i^2\rangle = \frac{1}{2}M\,\langle V^2\rangle + \frac{M+m}{M}\cdot\frac{A_-(\infty,\omega)}{n_i^+\,\bar{\xi}} \tag{21}$$

which is valid under the assumption that the neutral is stationary before the collision in the case of a small scattering angle using relation (22)

$$\frac{1}{2}m\,v^2 = \frac{1}{2}m\,v_i^2 - \frac{1}{2}MV^2 \tag{22}$$

Eq. (21) reduces with the aid of $m<v_{i}^2> = 3kT$ and $M<V^2> = 3kT$ to

$$kT_i = kT + \frac{M + m}{3M} \frac{q^2 E_1^2}{4m (\bar{E}^2 + \delta\omega^2)} \qquad (23)$$

This is the expression for the heated ion energy due to the power absorption from the rf field. Evaluating measurements of the half widths of ICR spectral lines this relation should be taken into account. Viehland and Mason [16] proposed an alternate technique for determining the half widths of ICR spectral lines, which is based on the relation (23), too.

Ion-Neutral Interactions and Evaluation of the Collision Frequencies

The Lennard-Jones potential is a reasonable approximation to the interaction between neutrals for the study of transport properties. For an ion-neutral potential, it is believed that the charge-induced quadrupole energy should be added to the Lennard-Jones potential. A model embodying these characteristics is a (12-6-4) potential, which can be written as

$$H_{im} = (\varepsilon/2) [(1+\gamma) (a/r)^{12} - 4\gamma(a/r)^6 - 3(1-\gamma) (a/r)^4] \qquad (24)$$

where ε and a are the depth and position of the potential minimum and γ is a dimensionless third parameter that measures the relative strengths of the r^{-6} and r^{-4} terms. Using the ion-neutral potential (24), the decoupling procedure of equation (11) is as follows. As already stated, we can convert the difference of the distribution functions W_{m1} into the difference of the velocities exchanging the role of statevectors for the last term in (10), namely

$$\sum_{m,1} <v>_m \hat{\bar{E}}_{m1} = \sum_{m,1} \{<v>_m - <v>_1\} \bar{\bar{E}}_{1m} \qquad (25)$$

because the symmetry relation $\bar{\xi}_{1m} = \bar{\xi}_{m1}$ is valid.

The quantity $\bar{\xi}_{1m}$ can be expressed with the wave packets (4) as

$$\bar{\xi}_{1m} = (2\pi/\hbar) \sum_{n,n} (J_n^2(\varkappa a)/n!e) (J_{n^-}^2(\varkappa a)/n!e) |<n|H_{im}|n^->|^2 \delta(\varepsilon - \varepsilon_{1m})$$

(26)

with the quantum state vectors $|n> \;(<r_\perp|n> = \psi_{\perp,n}(\vec{r}_\perp,\vec{r}_M)$ in (2).

Making use of the orthonormality of wave packets the expression (26)

turns out to be

$$\bar{\xi}_{1m} = (2\pi/\hbar) \sum_{n} <n|(H_{im})^2|n> (J_n^2(\varkappa a)/n!e) \; \delta (E - E_{1m})$$

(27)

For the evaluational purposes we express the relative distance between

an ion and a neutral in terms of coordinates of the cycloidal motion

center r_M: $\vec{r} = \vec{r} - \vec{r}_M + \vec{r}_M$. Further we make use of the relation

$(1 + 2(r_M/\rho)\cos\varphi + (r_M/\rho)^2)^{-2\upsilon} = \sum_{\alpha=0}^{\infty} C_{2\alpha}^{2\upsilon}(\cos\varphi)(r_M/\rho)^{2\alpha}$ in which the coeffi-

cients $C_{2\alpha}^{2\upsilon}(\cos\varphi)$ are the functions of Gegenbauer [17]. The angles be-

tween the vectors $\vec{r}_M$ and $\vec{\rho} = \vec{r} - \vec{r}_M$ can be $\varphi \rightarrow [0,2\pi]$ and should be averaged

on account of the isotropic nature of the collisions. Performing this pro-

cedure the ion-neutral potential squared can be represented as

$$H_{im}^2 = (\varepsilon^2/4) A_i a^{2\upsilon_i} \sum_{\alpha=0}^{\infty} C_{2\alpha}^{2\upsilon_i}(\cos\varphi) (r_M)^{2\alpha} \rho^{2\upsilon_i - 2\alpha}$$

(28)

where we have expressed, for the sake of convenience, the ion-neutral po-

tential squared applying a sort of dummy index summation rule:

$$\upsilon_1 = 12, \; \upsilon_2 = 9, \; \upsilon_3 = 8, \; \upsilon_4 = 6, \; \upsilon_5 = 5, \; \upsilon_6 = 4$$

and $A_1 = (1+\gamma)^2, \; A_2 = -8\gamma(1+\gamma), \; A_3 = -6(1-\gamma^2), \; A_4 = 16\gamma^2$

$A_5 = 24\gamma(1-\gamma)$ and $A_6 = 9(1-\gamma)^2$

After carrying out the integral calculations

$$<n|\rho^{-2\nu_i} {}^{-2\alpha}|n> = (b/2)^{\nu_i+\alpha} [n(n+1)\cdots\cdots(n-\nu_i-\alpha+1)]^{-1}$$

(29)

and converting the Landau quantum number into the cycloidal radius squa-

red $n \simeq (b/2)R^2$ for $[n(n-1)\ldots(n-\nu_i-\alpha+1)]^{-1} \simeq n^{-(\nu_i+\alpha+1)}$

and $(\Delta E)^{-1} \simeq (E/\sqrt{n})^{-1}$ which is obtained from the factor $\delta(E-E_{1m})$

we obtain finally

$$\bar{\xi}_{1m} = (2\pi/h) (\epsilon^2/4E_{ion})A_i(a/R)^{2\nu_i} (2/bR^2)^{1/2} \sum_{\alpha=0}^{\infty} C_{2\alpha}^{2\nu_i}(\cos\varphi) (r_M/R)^{2\alpha}$$

(30)

which is free from the Landau quantum numbers. Inserting (30) into (27)

and making use of orthonormality of wave packets the expression (27) also

becomes free from the Landau quantum numbers and the decoupling is per-

formed.

The collision frequency

$$\bar{\xi}_{1m} = (2\pi/\hbar) (\epsilon^2/4 E_{ion}) A_i (a/R)^{2\nu_i} (2/bR^2)^{1/2} \cdot$$

$$\cdot \sum_{\alpha=0}^{\infty} (-1)^{\alpha} \frac{\Gamma(2\nu_i+\alpha)}{\Gamma(2\nu_i)\alpha!} {}_2F_1(-\alpha,\alpha+2\nu_1;1/2,\cos^2\varphi) \times (\frac{r_M}{R})^{2\alpha} \quad (31)$$

contains the adjustable parameters ϵ, a and γ, which can be fitted through

the measurements of half widths of ICR spectral lines. The collision rate

depends on the rotation energy of ions and the ratios of position of po-

tential minimum to cycloidal radius and that of coordinates of center to

cycloidal radius. The inverse proportionality of collision rate to the

ion energy reflects that the fast ions undergo less collisions than the

slow ions.

Discussion

In this paper we have given a quantum theoretical description of ion cyclotron resonance collision broadening utilizing the minimized wave packet approach to the nearly classically behaving ion motions in ICR cells. The basic equation of an average ion motion in ICR cells, namely the phenomenological Langevin equation of motion is derived from the Neumann mean value of the Heisenberg velocity operator linearizing the statistical density operator with regard to the ion-neutral potentials. The expression for the power absorption from the rf fields can be obtained from the solution of the Langevin equation. In this way we can avoid many assumptions for the collision terms which constitute the principal difficulty for those who are concerned with the transport properties of matter. This theory is valid for electric fields of arbitrary strength.

Evaluating the collision rate for the specified ion-neutral potentials we determined the factors on which the collision rate depends. The theory described with regard to the energy dependence of collision rates leads to the result that the collision cross sections are inversely proportional to the velocity.

The minimized wave packet approach to the ion motions in the ICR cells offers the possibility to evaluate the various quantities concerning ICR experiments rigorously and opens a certain way to investigate the form of the ion-neutral potential and various chemical and physical properties of molecules availing oneself of collision dominated ICR experiments.

Acknowledgement

One of us (K.M.C.) wishes to thank the Akademie der Wissenschaften und der Literatur zu Mainz for financial support.

References

1. Hartmann H., Chung K.-M., and Baykut G. Jr: to be published

2. Viehland L.A., Mason E., and Whealton J.H.: J. Chem. Phys.
   62, 4715 (1975)

3. Hirshfelder J.O., Curtiss C.F., and Bird R.B.: Molecular Theory
   of Gases and Liquids Wiley, New York, 1964

4. McDaniel E.W., Mason E.A.: The Mobility and Diffusion of Ions
   in Gases Wiley, New York, 1973

5. Hartmann H., Lebert K.-H., and Wanczek K.-P.: Topics in current
   chemistry 43, 57 (1973)

6. Whealton J.-H., Mason E.A., and Robson R.E.: Phys. Rev. A9, 1017
   (1973)

7. Viehland L.A., Mason E.A.: Ann. of Phys. 91, 499 (1975)

8. Kihara T.: Rev. Mod. Phys.: 25, 944 (1953)

9. Beauchamp J.L.: J. Chem. Phys. 46, 1231 (1967)

10. Atkins, P.W., Clugston M.J.: Mol. Phys. 31, 603 (1976)

11. Hartmann H., Chung K.-M.: Theoretica Chim. Acta (Berl.) 45,
    137 (1977)

12. Chung K.-M., Mrowka B.: Z. Physik 259, 157 (1973)

13. Barut A.O.: Z. Naturforsch. 32a, 362 (1977)

14. Dirac P.A.M.: The Principles of Quantum Mechanics, 3rd ed.
    Oxford 1947

15. Buttrill, S.E.: J. Chem. Phys. 50, 4125 (1969)

16. Viehland L.A., Mason E.A.: Int. J. of Mass Spect. Ion Phys.,
    21, 111 (1976)

17. Magnus, W., Oberhettinger F.: Formeln und Sätze für die speziellen
    Funktionen der mathematischen Physik Springer, Berlin, Göttingen-
    Heidelberg, 1948

Improvement of the Electric Potential

in an Ion Cyclotron Resonance Cell

Jungji Urakawa, Hiromi Shibata and Masao Inoue
Department of Engineering Physics,
The University of Electro-Communications
1-5-1, Chofugaoka, Chofu-shi, 182 Japan

Introduction

Ion cyclotron resonance (ICR) spectrometry has become a useful tool
for studying ion chemistry in the gas phase. It utilizes the motion of
ions in electric and magnetic fields. In the conventional ICR spectrometer,
various RF and DC electric fields are produced by a number of electrodes
which compose an ICR cell. The principal functions of these fields are to
trap ions inside the cell, to drive ions by cycloidal motion in the direction perpendicular to the static electric and magnetic fields and to make
resonant the ions which meet the cyclotron resonance condition. It is desirable that these fields are as uniform as possible in order to obtain
quantitative information from the ICR signals. However, in most ICR cells
the electric fields are quite inhomogeneous and this frequently makes the
interpretation of experimental results difficult.

In this paper we propose a new type of ICR cell which has improved
electric fields. First, we discuss briefly the conventional ICR cell which
is widely used in most ICR experiments and then the modifications of this
cell which were designed to improve the electric fields. Secondly, theoretical comparisons are made between the conventional and proposed cells and
finally, effects of ICR cell shape on various cell characteristics are
discussed.

The Conventional ICR Cell and Its Modifications

Fig. 1 shows the cross sectional view of the conventional "square" cell[1-3] with equipotential patterns. The cell possesses a pair of drift electrodes which, in combination with a magnetic field parallel to these electrodes, cause ions to migrate in the direction perpendicular to the

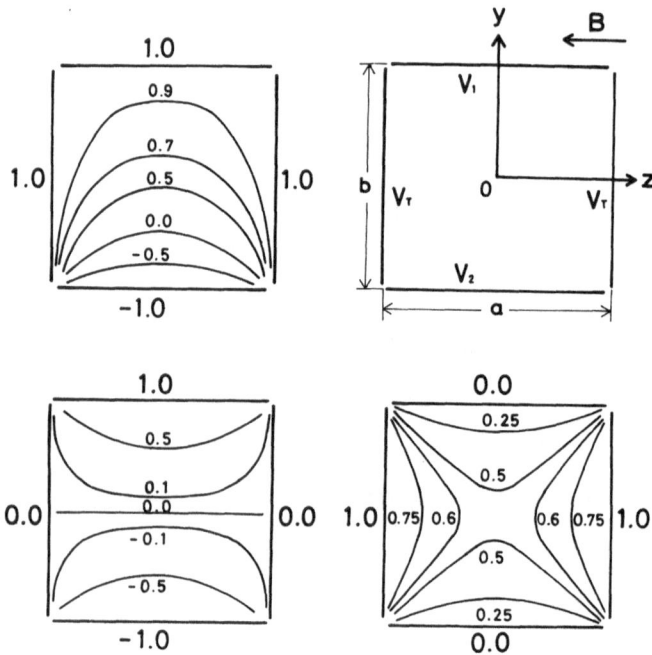

Fig. 1. Cross sections of a conventional ICR square cell with equipotential lines. The pattern in the upper left-hand side shows the potentials typically employed in the ICR experiments. The patterns in the lower left-hand and right-hand sides show two extreme conditions where no trapping or no drift voltage is applied. The dimensions of the cell, the direction of the magnetic field and the coordinates are shown in the upper right-hand side in the Figure.

direction of the magnetic field and a trapping voltage of the same polari-

ty as that of the ions is applied to these electrodes. The equipotential

pattern in the upper half of Fig. 1 is typically employed for ICR experi-

ments where ions are trapped over the entire cross section of the cell.

The pattern in the lower left- and right-hand sides represent two extreme

conditions for which no trapping voltage or no drift voltage is applied

to the corresponding electrodes. It can be seen that these electrodes re-

sult in electric fields inside the cell which are quite complicated and

are far from the uniform fields which are generally assumed when descri-

bing ion motion in the cell. Therefore in a real ICR cell, the motion of

ions is also quite complicated. It can de described by three types of su-

perimposed motions:[4] (1) the drift motion in the direction perpendicular

to the plane, (2) harmonic oscillation in the trapping potential well, and

(3) quasi-cyclotron motion about the drifting center in the plane perpen-

dicular to the magnetic field. These motions vary with the position of

ions in the cell due to the inhomogeneity of the electric field. In addi-

tion, since the ions are distributed in the cell, the resultant ICR signals

also depend on the spatial distribution of ions.[5] The effect of the field

distribution manifests itself in many ways in ICR experiments, and include

shifts in the resonance magnetic field,[2,6] asymmetry of the lineshapes,[7]

low mass resolution due to the spread of resonance frequencies,[8] and poor

resolution in trapping ejection experiments.[6]

In order to reduce the inhomogeneity of the electric field, some modi-

fications of the original "square" cell have been attempted. One modifica-

tion is to reduce the spacing between the drift electrodes, thus attenua-

ting the effect of the trapping potential at the center of the cell.[8]

This "flat" cell configuration has been widely employed to obtain quantitative results. The second example is the four-section cell reported by Clow and Futrell.[9]   The reaction and analyzer region of the cell are basically of "square" cell dimensions but the outer one fourth of the drift plates are bent 50^0 from the horizontal towards the center of the cell. The cell was constructed to provide a better ion transmission which results from the more uniform electric field in the center of the cell. The third example is the trapped ion analyzer cell described by McIver et al.[10] Their cell is composed of hyperbolic electrodes instead of flat plates, since the hyperbolic electrodes generate a more uniform quadrupole electric field in the center of the cell.

In order to improve the electric field, we would like to propose another type of cell which has Venetian-blind type trapping electrodes instead of the conventional flat plates. In the upper half of Fig. 2 equipotential patterns obtained with such a cell are illustrated. The trapping electrodes consist of a set of equally spaced electrodes between the upper and lower drift plates so that the potentials of these electrodes can be controlled independently. The voltage on successive electrodes of the Venetian-blind were incremented in equal steps such that the voltage difference across the Venetian-blind was equal to the potential difference between upper and lower drift electrodes. At the same time, the center electrode of the Venetian-blind was set at some specified trapping bias voltage. In this way, when the trapping bias voltage becomes zero, the equipotential lines become parallel to the drift electrodes. This is the potential configuration of a parallel-plate capacitor and corresponds to the simplest electric field which is often used when describing ICR expe-

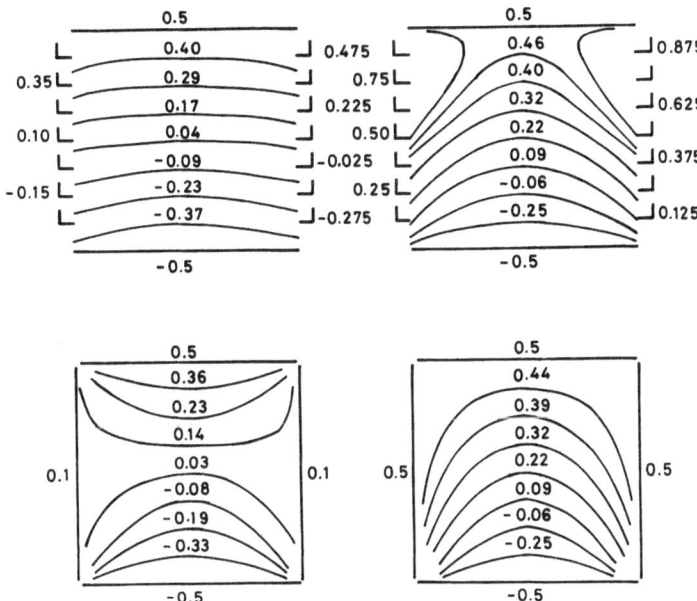

Fig. 2. Comparison of equipotentials between the proposed and the conventional cells. The patterns in the upper left-hand and in the lower right-hand sides show the electric fields most appropriate to the two types of cells. As seen in the upper right-hand side, when the trapping bias voltage of the proposed cell is raised to 0.5 V, the equipotential pattern becomes complicated and not appropriate to this configuration of the elctrodes. The pattern in the lower left-hand side is obtained with the conventional cell, when 0.1 V is applied to the trapping electrodes. In this case ions in the upper half of the cell cannot be trapped but are instead ejected from the cell.

riments. For comparison equipotential patterns of the conventional "square" cell are shown in the lower half of the figure where the trapping voltages are chosen to be the same as the trapping bias voltages of the proposed cell. The pattern of the lower right-hand side is the typical electric field employed for most ICR experiments. For the proposed cell two equipotential patterns are illustrated where bias voltages are + 0.1 and + 0.5 V for the same drift voltages of ± 0.5 V. It can be seen that the trapping bias voltage of + 0.1 V produces an electric field quite uniform over the cross section of the cell with a trapping potential well for positive ions of about 0.1 V which is enough to trap thermal ions in the cell. However, the trapping voltage which is typically used for the conventional cell is not appropriate to the proposed cell. The pattern in the upper right-hand side illustrates such a case where a bias voltage of + 0.5 V makes the electric field quite complicated. In the case of the conventional cell, the effective space where ions are trapped narrows as the trapping voltage decreases. The pattern in the lower left-hand side illustrates equipotentials obtained with a trapping voltage of + 0.1 V. In this case positive ions produced in the space above the equipotential line of + 0.1 V cannot be trapped but are accelerated towards the trapping plates.

Theoretical Comparison between the Conventional and Proposed Cells

Electric Potential and Equation of Motion of the Ions

We first derive the electric potential of the proposed cell. If we increase the number of the trapping electrodes infinitely, the trapping

electrodes can be replaced by a flat plate on which the potential conti-
nuously varies. Therefore for the rectangular cross section and the coordinate system shown in Fig. 3, we have chosen the boundary condition
as follows:

$$V(\tfrac{b}{2}, z) = V_1, \quad V(-\tfrac{b}{2}, z) = V_2 \quad \text{for} \quad -\tfrac{a}{2} \leq z \leq \tfrac{a}{2}, \tag{1}$$

$$V(y, \tfrac{a}{2}) = V_T, \quad V(y, -\tfrac{a}{2}) = V_T \quad \text{for} \quad -\tfrac{b}{2} \leq y \leq \tfrac{b}{2}, \tag{2}$$

where $\quad V_T = \tfrac{1}{b}(V_1-V_2)y + \tfrac{1}{2}(V_1+V_2) + V_t,$ \hfill (3)

a and b are the width and height of the cell. V_1 and V_2 are the potentials
applied to the upper and lower drift electrodes. V_T is the trapping potential which is applied to the trapping electrodes at $z = \pm\tfrac{a}{2}$. It contains
y coordinate in the first term and the trapping bias voltage V_t in the
third term.

The solution of the Laplace equation for these boundary conditions
gives

$$V = \tfrac{1}{b}(V_1-V_2)y + \tfrac{1}{2}(V_1+V_2) + V_t - \frac{4V_t}{\pi} \sum_{k=0}^{\infty} \frac{(-)^k}{2k+1} \frac{\cosh\{(2k+1)(\pi y/a)\}}{\cosh\{(2k+1)(\pi b/a)\}}$$

$$x \cos\{(2k+1)(\pi z/a)\} \cdot \tag{4}$$

This is an equation of a two-dimensional potential as a function of y and
z. The coordinate y appears in the first term and in the hyperbolic cosine term, and z is in the cosine term.

We can expand the hyperbolic cosine and cosine term in power series
to obtain the exact electric field (see a recent paper by Knott and
Riggin[4] on the analysis of ion motion in a conventional rectangular cell):

$$E_y(y,z) = \sum_{m,n=0}^{\infty} e'_{m,n} y'^m z'^{2n}, \tag{5}$$

$$E_z(y,z) = -\sum_{m,n=0}^{\infty} \{(m+1)/(2n+1)\} e'_{m+1,n} y'^m z'^{2n+1}, \tag{6}$$

where $\quad y' = \dfrac{y}{a}, \quad z' = \dfrac{z}{a},$

$$e'_{m,n}(m:even) = 0 \quad \text{except for} \quad m = 0 \tag{7}$$

$$e'_{0,0} = -2V_d/b \quad \text{where} \quad V_d = \tfrac{1}{2}(V_1 - V_2), \tag{8}$$

$$e'_{m,n}(m:odd) = \frac{4V_t}{a} \frac{(-)^n}{m!(2n)!} \sum_{k=0}^{\infty} \frac{(-)^k \{(2k+1)\pi\}^{m+2n}}{\cosh\{(2k+1)(\pi b/2a)\}}, \tag{9}$$

E_y and E_z are y and z components of the electric field. The prime is used for the coefficients $e'_{m,n}$ because they are different from those employed by Knott and Riggin by a factor of $a^{-(m+n)}$ so that $e'_{m,n}$ have the dimensions of electric field strength. V_d, which is equal to half of the potential diffe-rence between the drift electrodes, should appear in the $e'_{m,n}$ for even va-lues of m. However, for the proposed cell they become zero except for $e'_{0,0}$ which is equal to $-2V_d/b$ (for a conventional cell, $e'_{m,n}$ for even values of m are not zero, but have finite values). The trapping bias voltage V_t ap-pears in $e'_{m,n}$ for odd values of m.

The equation of motion for an isolated ion of mass M and charge q is given by

$$M \vec{v} = q(\vec{E} + \vec{v} \times \vec{B}), \tag{10}$$

where $\vec{v}$ is the velocity of the ion and $\vec{B}$ is the magnetic field strength which is along the z direction. The equation may be written in component form:

$$\ddot{x} = -\omega_c \dot{y}$$
$$\ddot{y} = \omega_c \dot{x} + \frac{q}{M} E_y(y,z) \tag{11}$$
$$\ddot{z} = \frac{q}{M} E_z(y,z)$$

where $\omega_c = qB/M$.

The equation of motion of an ion in a conventional rectangular cell was first solved by Beauchamp et al.[11] They approximated the electric field as the sum of an electric field of a parallel-plate capacitor and a quadrupole field, the former approximating the drift field and the latter the trapping field. Because of the approximations made with the electric field, their results are only applicable to the ions very near the center of the cell. Another solution of the equation for the conventional rectangular cell, and a more precise field, has been worked out recently by Knott and Riggin.[4] They took terms up to the fourth order in the y-coordinate in a Taylor expansion of the exact electric field. With this potential the motion of ions in the x-y plane and along the z-axis are coupled and the equations of motion becomes nonlinear and difficult to solve. Nevertheless they determined a procedure for obtaining approximate decoupled equation which may be solved using Weierstrass elliptic functions. We followed their procedure to obtain a solution of the equation of motion for the proposed cell. We will not reproduce here the mathematical procedure but only present some results obtained from the solution.

Shift in the Resonance Magnetic Field

The shift in the resonance magnetic field ΔB as a function of initial position (x_0, y_0, z_0) and initial velocity $(\dot{x}_0, \dot{y}_0, \dot{z}_0)$ of ion in the cell is expressed as

$$\Delta B = \frac{1}{a\omega_c} (e_1' + 6e_3'Y_0'^2) \qquad \text{(Tesla)}, \tag{12}$$

where

$$e_m' = (\frac{1}{m+1}) \sum_{n=0}^{\infty} e_{m,n}' z'^{2n}, \tag{13}$$

and

$$Y_0' = \frac{1}{a}\{y_0+(\dot{x}_0/\omega_c)\}, \tag{14}$$

Y_0' is a variable related to the initial y coordinate and x·component of the initial ion velocity. e_m' is function of the z-coordinate and e_m' for odd values of m contains the trapping bias voltage V_t.

For comparison we quote here the shift in the resonance magnetic field of the conventional rectangular cell given by Knott and Riggin:[4]

$$\Delta B = \frac{1}{a\omega_c}(e_1' + 3e_2'Y_0' + 6e_3'Y_0'^2) \qquad \text{(Tesla)}. \tag{15}$$

For the conventional rectangular cell $e_{m,n}'$ for even values of m is not zero but given by

$$e_{m,n}'(\text{m:even}) = \frac{4V_d}{a} \frac{(-)^n}{m!(2n)!} \sum_{k=0}^{\infty} \frac{(-)^k\{(2k+1)\pi\}^{m+2n}}{\sinh\{(2k+1)(\pi b/2a)\}} \tag{16}$$

$e_{m,n}'$ for odd values of m has the same expression as for the proposed cell, but V_t which appears in the equation has a different meaning:

$$e_{m,n}'(\text{m:odd}) = \frac{4V_t}{a} \frac{(-)^n}{m!(2n)!} \sum_{k=0}^{\infty} \frac{(-)^k\{(2k+1)\pi\}^{m+2n}}{\cosh\{(2k+1)(\pi b/2a)\}}, \tag{9}$$

where

$$V_t = V_T - \frac{1}{2}(V_1 + V_2), \tag{17}$$

V_T is the trapping voltage applied to the electrodes at $z = \pm \frac{a}{2}$. It is apparent from Eqs. 12 and 15 that the shift of the improved cell depends only on the trapping bias voltage to the first approximation, while in the

case of the conventional cell the shift depends not only on the trapping
voltage but also on the drift voltage.

Figs. 3 and 4 show contour maps of the shift in the resonance field
calculated from Eqs. 12 and 15 for the two types of cells both having
a height to width ratio b/a of 0.56. The drift voltages are chosen to
be $\pm$ 0.5 V. The trapping bias voltage for the proposed cell and the trapping voltage for the conventional cell are + 0.1 and + 0.5 V respectively.
The values are given for the argon ion (M/q = 40) at an observing frequency of 297 kHZ and a magnetic field strength of 0.78 Tesla. The maps only
show the central region of the cell. The trapping plates are placed at
z/a = + 1/2 and the drift plates are at y_0/a = + 1/3.6. From Figs. 3 and
4, it can be seen that the electric field of the conventional cell is
quite distorted, especially in the lower part of the cell. In the region
shown the shift varies from 6.2×10^{-4} to 35.0×10^{-4} Tesla which correspond
to the shift in mass $\Delta(M/q)$ from 0.032 to 0.181. The shift of the proposed cell is symmetric about the z-axis and varies from 1.2×10^{-4} to
3.8×10^{-4} Tesla corresponding to $\Delta(M/q)$ from 0.006 to 0.020. Thus the variation is only one tenth that of the conventional cell.

Distribution of Ion Drift Velocity in the Cell

An ion in the cell drifts in the x direction in cycloidal paths by
the combined effect of the electric and magnetic fields. The net drift
velocity v_D is expressed as

$$v_D = -e_y(y_0)/B \qquad \text{(m/sec)}, \qquad (18)$$

where $e_y(y_0)$ is the electric field in the y-direction as a function of the
initial y coordinate of the ion. Using the same coefficients e_m' as for the

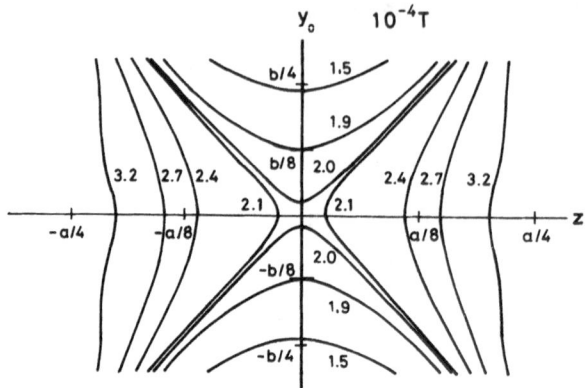

Fig. 3. Contours of ΔB = const. in the cross section of the proposed cell.
The lines are calculated using Eq. 12 with $\dot{x}_0$ = 0. The cell dimensions are
a = 0.025 m and b = 0.014 m. V_t = 0.1 V and V_d = 0.5 V. The mass of ion is
40 a.m.u. and B = 0.780 Tesla.

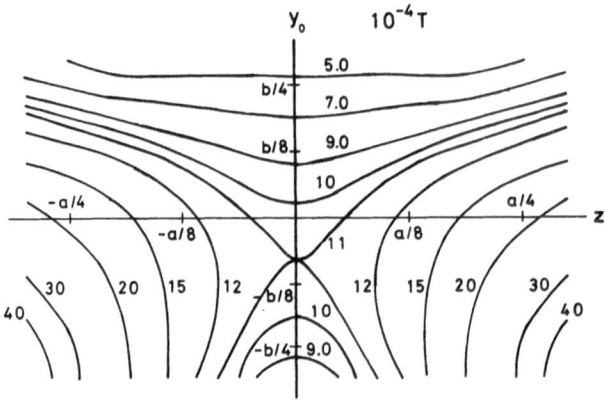

Fig. 4. Contours of ΔB = const. for the conventional rectangular cell
calculated from Eq. 15 . Conditions are the same for Fig. 3
except V_t = 0.5 V.

proposed cell by

$$e_y(y_0) = e'_{0,0} + 2e'_1 y'_0 + 4e'_3 y'^3_0 ,$$ (19)

and for the conventional rectangular cell by

$$e_y(y_0) = e'_0 + 2e'_1 y'_0 + 3e'_2 y'^2_0 + 4e'_3 y'^3_0 ,$$ (20)

It is seen that the expression for the proposed cell differs from that for the conventional cell in the first term and, in addition, does not have the term containing the coefficient e'_2.

Numerical results on the distribution of the drift velocities in the cell were obtained for the same condition as for the shift in the resonance magnetic field. Figs. 5 and 6 show contour maps of the drift velocity in the cross section of the two cells. In the case of the proposed cell the map is nearly symmetric about the z-axis and the velocity varies from 88 to 102 m/sec while in the conventional cell the spread is in the range from 50 to 120 m/sec which is about five times greater than that of the proposed cell.

Variation of the Trap Oscillation Frequencies

Ions in the cell oscillate in the z-direction due to the potential barrier produced by the different electrodes. The frequency of the trap oscillation ω_T is given by

$$\omega_T^2 = \frac{q}{Ma}(e'_{1,0} + 3e'_{3,0} y'^2_0)$$ (21)

for the proposed cell, and by

$$\omega_T^2 = \frac{q}{Ma}(e'_{1,0} + 2e'_{2,0} y'_0 + 3e'_{3,0} y'^2_0)$$ (22)

for the rectangular cell. Since ω_T^2 of the proposed cell contains only the coefficents $e'_{m,n}$ having odd values of m the trap oscillation frequency

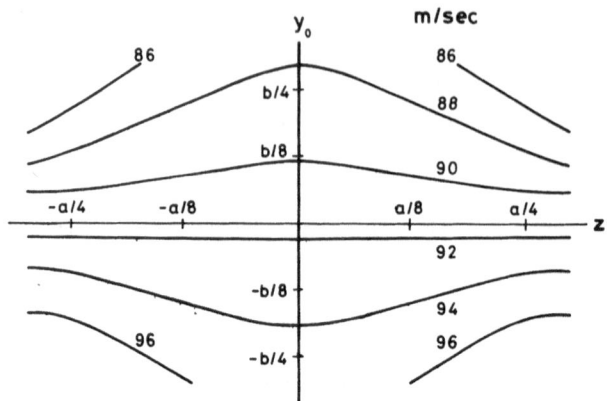

Fig. 5. Contours of v_D = const. for the proposed cell calculated from Eqs. 18 and 19 . Conditions are the same as for Fig. 3.

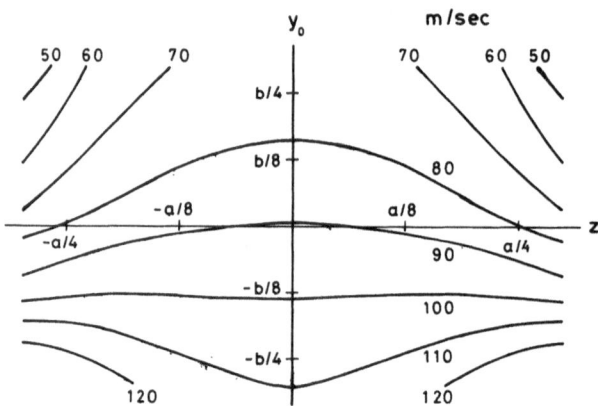

Fig. 6. Contours of v_D = const. for the conventional rectangular cell calculated from Eqs. 18 and 20 . Conditions are the same as for Fig. 4.

depends only on the trapping bias voltage to the first approximation. On the other hand, for the conventional cell it depends not only on the trapping voltage but also on the drift voltage.

Variation of the trap oscillation frequency for argon ions as a function of the y coordinate is shown in Fig.7. The solid curve represents the variation for the proposed cell and the broken curve is for the conventional cell. For comparison the trapping bias voltage of the proposed cell was chosen to be + 0.5 V the same as that of the conventional cell. The trap oscillation frequency is converted to an equivalent M/q value at y_0 = 0 and is indicated in the Figure.

Effects of Cell Dimension on the Various Characteristics

It is interesting to note effects of width to height ratio b/a on the various characteristics of the cell. Using the same formulae as for the cell of b/a = 0.56, we carried out calculations of the shift in the resonance magnetic field, drift velocity and the trap oscillation frequency for cells having b/a = 1.00 and 2.00. In these calculations potentials applied to the various electrodes were kept the same as for the case where b/a = 0.56.

Fig. 8 shows the shift in the resonance magnetic field ΔB, calculated for an area of one fourth of the cross section of the improved cell. Since the electric field strength in the cross section varies with the ratio of b/a and is proportional to the potentials applied to the different electrodes, the absolute values of ΔB should not be compared between cells having different ratios. Therefore, to illustrate the effect of b/a, we have expressed deviations of the shift from the value at the center

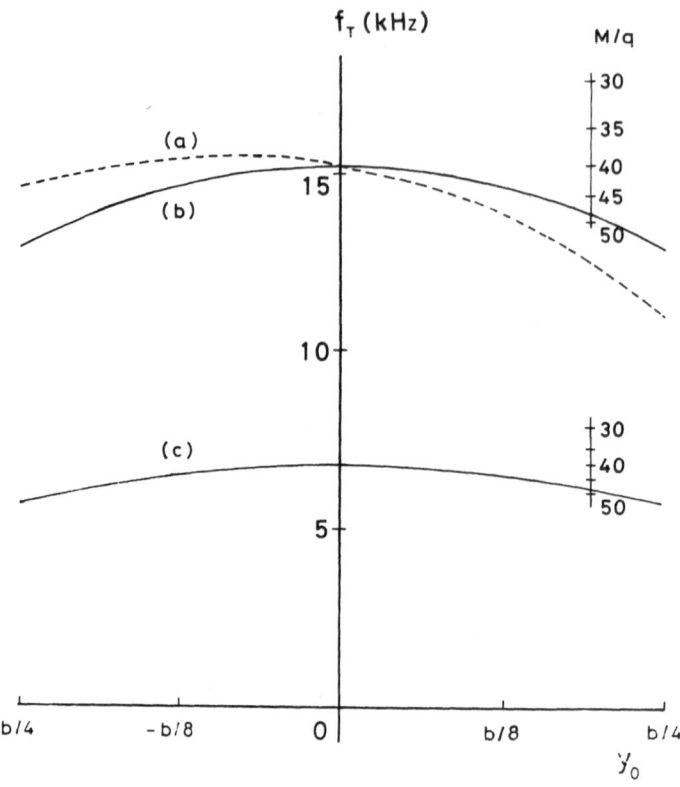

Fig. 7. The variation of the oscillation frequency $f_T = \omega_T/2\pi$ for the ion of M/q = 40 in the trapping field as a function of the y coordinate of the ion. The solid curves (b) and (c) are calculated for the proposed cell using Eq. 21 and the broken curve (a) is for the conventional rectangular cell calculated from Eq. 22 . V_t is 0.5 V for (a) and (b), and 0.1 V for (c). The other conditions are the same as for Fig. 3. The ejection frequency is converted to equivalent M/q values at $y_0' = 0$ and indicated in the Figure.

of the cell in percentages and have written these in the left half of
each Figure. It seems that in the case of the improved cell the ratio
b/a = 1.00 gives more uniform shift than the other two ratios. Fig. 9
illustrates results of the same calculation for the conventional cells.
By comparison with Fig. 8 it can be seen that values of ΔB are greater
than those of the improved cell for each ratio. As for the case of the

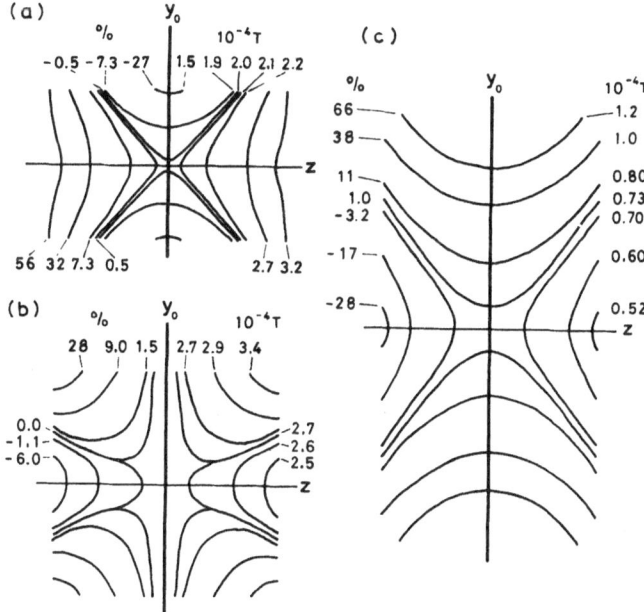

Fig.8. The effect of b/a ratio on the shift in the resonance magnetic
field ΔB for the proposed cells calculated from Eq. 12 with $\dot{x}_0$ = 0.
V_d = 0.5 V and V_t = 0.1 V for all cases. Cell dimensions are a = 0.025 m,
b^d= 0.014 m and b/a = 0.56 for the contour map (a), a = 0.021 m, b = 0.021 m
and b/a = 1.00 for (b), and a = 0.020 m, b = 0.040 m and b/a = 2.00 for (c).
The ion mass is 40 a.m.u. and the magnetic field strength is 0.780 Tesla.
To illustrate the effect of b/a ratio on the uniformity of ΔB, variation
of ΔB from the value at the center of the cell is expressed in percentages
in the left half of each Figure.

improved cell the variation of the shift is more uniform with the ratio

b/a = 1.00.

Fig. 10 compares contour maps of the drift velocity v_D calculated

for improved cells having three different ratios. Deviations from the

values of v_D at the center of the cell are written in the left half of

each map. The effect of the ratio b/a on the variation of the drift ve-

locity seems small with the improved cell. Fig. 11 shows the correspond-

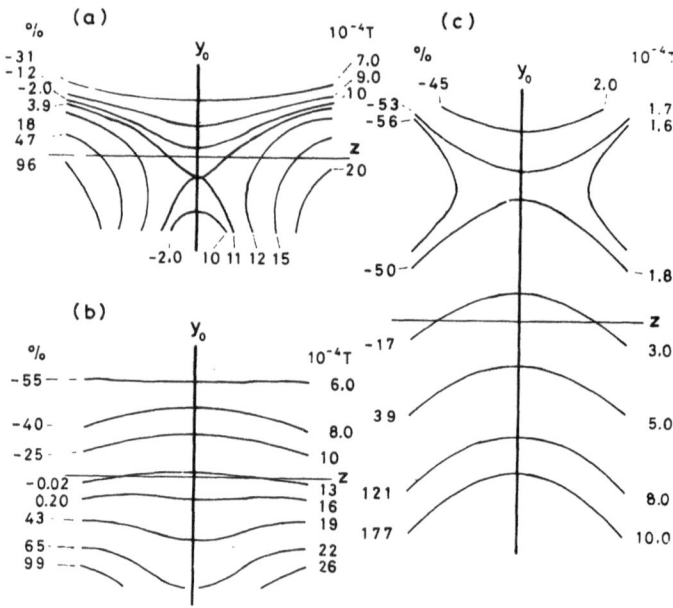

Fig. 9. The effect of the b/a ratio on the distribution of ΔB for the con-

ventional rectangular cell. The contours are calculated from Eq.

15 . Dimensions of cells and other conditions are the same as

for Fig. 8 except V_t = 0.5 V.

ing maps calculated with the conventional cells. Among the three cases

the cell having b/a = 0.56 is the best with respect to the uniformity of

the drift velocity distribution as was anticipated with the "flat" cell.

Dependence of the variation of the trap oscillation frequency f_T on

the ratio of b/a is shown in Fig. 12. In this figure the abscissa repre-

sents the y coordinate of the ion and the ordinate is the trap oscilla-

tion frequency of the ion in the z direction. The broken lines show the

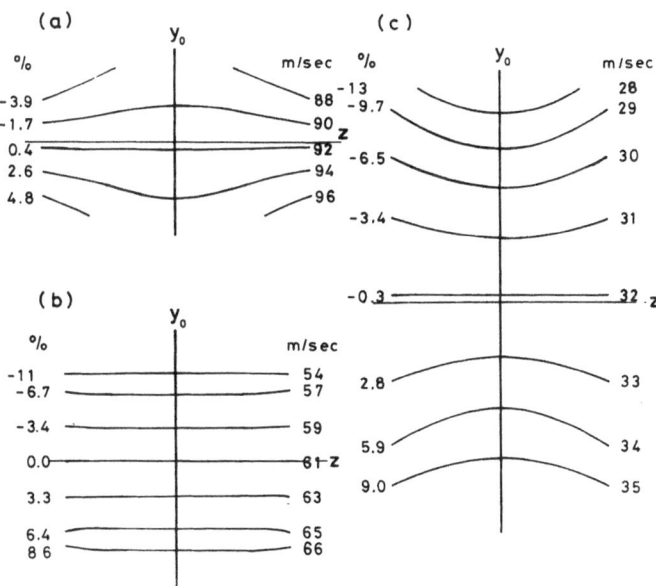

Fig. 10. The effect of the b/a ratio on v_D for the proposed cell calculated
from Eqs. 18 and 19. Conditions are the same as for Fig. 8.

variation of the trap oscillation frequency calculated for the conventional cells having b/a = 0.56, 1.00 and 2.00 with the same boundary conditions as above. The effect of b/a on the variation of f_T with the position of y_0 was also calculated for the improved cell and shown in Fig. 12 as the solid curves. As was the same with Fig.7, a trapping bias voltage of 0.5 V instead of 0.1 V was used for the improved cells to make

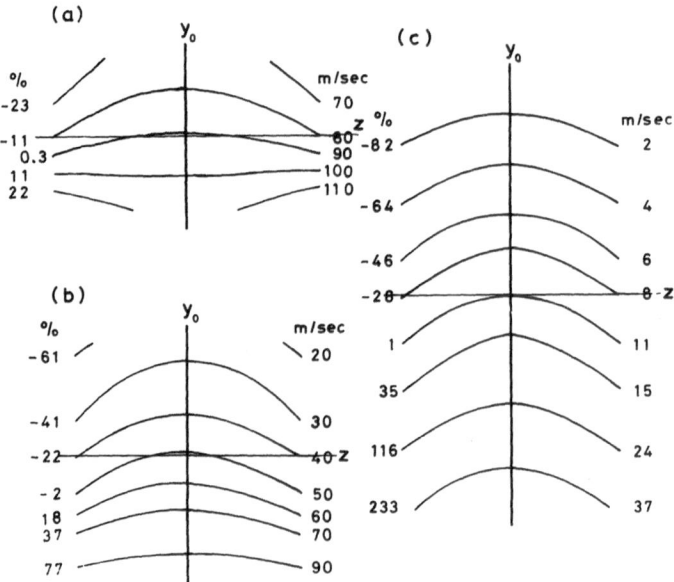

Fig. 11. The effect of the b/a ratio on v_D for the conventional rectangular cell calculated from Eqs. 18 and 20. Conditions are the same as for Fig. 9.

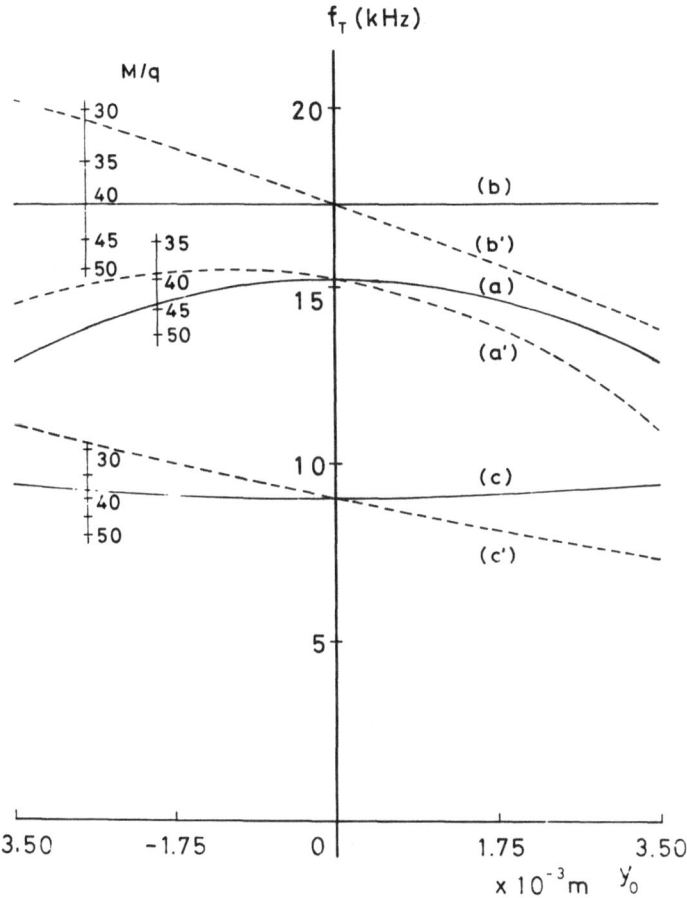

Fig. 12. The effect of the b/a ratio on the variation of the ejection frequency f_T as a function of the initial y coordinate of the ion. The solid
curves represent the proposed cells and the broken curves are for the conventional rectangular cells. (a) and (a') are for b/a = 0.56, (b) and (b')
are for b/a = 1.00, and (c) and (c') are for b/a = 2.00. The curves are calculated using Eqs. (21) and (22). V_d = V_t = 0.5 V for both types of cells.
The frequencies of the ejecting RF field are converted to M/q values at
y_0 = 0 and are indicated in the Figure.

f_T coincide with those of the conventional cells at $y_0/a = 0$. It can be seen that in the case of the improved cell the variation of the trap oscillation with y-coordinate is symmetric about the center of the cell and is much smaller than that of the conventional cell with each b/a, but the most interesting point is that with b/a = 1.00 f_T remains constant regardless of the trapping bias voltage (within the limit of y_0 calculated).

We carried out numerical calculations only in the central one-fourth area of the cross section of the cell because of the various approximations made in finding the solution of the equation of motion of the ions. Including terms up to the fourth order in the power expansion with respect to y' and z', the numerical values calculated are very exact near the center of the cell, however, they become less exact when the ion is away from the center of the cell.

Errors in ΔB and v_D for the conventional cell are estimated as follows: at $y_0'=\pm\frac{1}{4}$ and $z'=\pm\frac{1}{4}$, 20%, 6% and 30% of the calculated values for b/a=0.56, 1.00 and 2.00 repsectively, and at $y_0'=\pm\frac{1}{8}$ and $z'=\pm\frac{1}{8}$, 6%, 5% and 10% for b/a =0.56, 1.00 and 2.00 respectively. Errors in f_T calculated at $z'=\pm\frac{1}{4}$ are 30% for b/a=0.56 and $y_0'=\frac{1}{4}$, 7% for b/a= 1.00 and $y_0'=\frac{1}{8}$, and 15% for b/a=2.00 and $y_0'=\frac{1}{8}$.

Errors for the improved cells are estimated to be less than a half of these values for the conventional cells.

Discussion and Conclusion

Although the theoretical results mentioned here are not sufficient to estimate the performance of a real ICR cell, it can be expected that the proposed cell may have several advantages over conventional cells.

The uniformity of the electric field makes it possible to enlarge the effective volume where ions can be trapped in the cell and consequently interaction between ions become less serious. As for the trapping potential, it can be made as small as the kinetic energy of ion, and by reducing the trapping bias voltage the electric field approaches that of a parallel plate capacitor and the frequency of ion resonance approches that are given by the exact cyclotron resonance condition. Thus it should be possible to determine precise M/q values by extrapolating the trapping bias voltage to zero.

In the conventional cell the ion drift field (y component of electric field in the y-z plane) depends on both the drift and trapping voltages applied to the electrodes, so that the drift voltage cannot be lowered beyond a certain value when ions are trapped. With regard to the proposed cell, the trapping field is independent of the drift voltage so that the drift voltage can be loweredwithout a loss of ions. Since the resolution of ICR absorption lineshape is determined by the residence time of ions in the cell, higher resolution may be obtained by improvement of the electric fields.

We found that with the cell corresponding to b/a = 1.00 the frequency resolution does not vary with the y coordinate of the ion within the limits of the calculation, so the proposed cell may have much higher mass resolution when ions are ejected in the z direction. In addition, because of

the low trapping voltage, only a small RF field is sufficient to eject
ions out of the cell. Therefore, the efficiency of ejection may be much
better than the ordinary cells.

It should be noted that the theoretical results derived for the proposed cell are not directly related to the signals which would be obtained
in practice. We would like to discuss some of the factors which are not
mentioned in the preceding sections that seem to affect real ICR lineshapes. Firstly, this paper treats only the motion of an isolated ion in the
cell, but in practice, many ions are in motion in the cell and resultant
ICR signals appear as an average of an ensemble of ions distributed in
the cell. Therefore, to derive the ICR lineshape theoretically, the distribution of ions must be determined and the appropriate averages made. Secondly , in this paper our discussion of the effects of the potential is limited to the cross section of the cell (y-z plane). In deducing the performance of an actual cell, however, other minor effects of the potential
must be considered. One of them is the variations of the potential in the
direction of ion drift (x-direction). Since different drift voltages are
often applied to the different regions of the cell, equipotential lines
may not be uniform in the direction of ion drift and the position of an ion
in the y-z plane may vary in passing from one to the other. An inhomogeneity of the electric field in the x-direction may also be formed near
the end of the cell by the collection electrode which is usually held near
ground in order to collect ions passing through the cell. In addition,
field distortion may also be caused by the space charge of the ionizing
electron beam as well as the ions trapped in the cell. Finally, in this
paper the trapping electrodes of the proposed cell are treated ideally

as flat electrodes where potentials vary continously, however in a practical design they would be replaced by a limited set of electrodes on which
potentials are applied stepwise from one electrode to the other and this
will produce some disturbance in the uniformity of the electric field
near the trapping electrodes.

In spite of the complex structure of the proposed cell the above
features motivated us to design and construct several cells of this type.
Preliminary experiments are now being carried out to test their performance.

Acknowledgement

This research was supported by the Nishina Memorial Foundation.
The authors wish to thank Prof. S. Fujiwara of the University of Tokyo
for his interest and encouragement of this work, and Dr. E. Parks of
Argonne National Laboratory for reading the manuscript.

References

1. J. D.Baldeschwieler, Science 159, 263 (1968)

2. J. L. Beauchamp, Ph. D. Thesis, Harvard University, Cambridge, Massa-
   chusetts, 1967

3. J. L. Beauchamp, J. Chem. Phys. 46, 1231 (1967)

4. T. F. Knott and M. Riggin, Can. J. Phys. 52, 426 (1974)

5. M. Bloom and M. Riggin, Can. J. Phys. 52, 436 (1974)

6. J. L. Beauchamp and T. Armstrong, Rev. Sci. Instrum. 40, 123 (1969)

7. M. Riggin, Int. J. Mass Spectrom Ion Phys. 22, 35 (1976)

8. S. E. Buttrill, Jr., J. Chem. Phys. 50, 4125 (1969)

9. R. P. Clow and J. H. Futrell, Int. J. Mass Spectrom. Ion Phys.
   8, 119 (1972)

10. R. T. McIver, Jr., E. B. Ledford Jr. and J. S. Miller, Anal. Chem.
    47, 692 (1975)

Thermodynamic Information from Ion-Molecule Equilibrium Constant Determinations

Sharon G. Lias
Institute for Materials Research
National Bureau of Standards
Washington, DC 20234

Introduction

According to the principles of thermodynamics, the equilibrium constant of any chemical equilibrium gives information about the thermochemistry of the chemical reactions in the forward and reverse directions:

$$-RT\ln K_{eq} = \Delta G^{\circ} = \Delta H^{\circ} - T\Delta S^{\circ} \qquad\qquad I$$

Equation I has been applied extensively to the derivation of thermochemical information, particularly in the liquid phase.

After the beginnings of the systematic study in the gas phase of the chemistry of ions in the mid-1950's, some ten years elapsed before the thermodynamic principles summarized in equation I were applied to the investigation of ion thermochemistry. Such studies, which permit the direct determination of such chemical properties of molecules as intrinsic acidities or basicities in the absence of solvent effects are invaluable for a fundamental understanding of chemistry.

The earliest systematic investigations of ion equilibria were concerned mainly with the thermochemistry of solvation itself[1]. By examining equilibria such as:

$$A^{+} + B \rightleftharpoons (A\cdot B)^{+} \qquad\qquad (1)$$

researchers were able to measure free energies, enthalpies, and entropies

of solvation, and even elucidate structures of solvated ions and mechanisms of solvation in a detailed manner not possible in the liquid
phase. More recently, attention has shifted to an examination of equilibria involving chemical reactions which are bimolecular in both
directions. This trend began in 1971, when Bowers, Aue, Webb, and
McIver[2] suggested that the study of proton transfer equilibria:

$$AH^+ + B \rightleftharpoons BH^+ + A \qquad (2)$$

would permit a more accurate evaluation of the thermochemistry of proton
transfer reactions and the relative heats of formation of protonated
molecules than the appeareance potential or so-called "bracketing"
techniques then in vogue would allow. It was an idea whose time had
come, for during the same month that this suggestion appeared in the
literature, three papers concerned with proton transfer equilibria -
one using ion cyclotron resonance spectroscopy[3] (ICR), one flowing
afterglow[4], and one, high pressure mass spectrometry[5] - were submitted
for publication. At this writing, well over 50 studies concerned with
the examination of such bimolecular ion-molecule equilibria in the gas
phase have appeared[2-69]. Although about 70% of these investigations were
concerned with proton transfer reactions (reaction 3)[2-49], the approach
has also been applied to examinations of the thermochemistry of
halide transfer[50-53]:

$$A^+ + BX \rightleftharpoons AX + B^+ \qquad (3)$$

(where X is F, Cl, Br, or I, and A^+ and B^+ are carbonium ions), of charge
transfer[54-62]:

$$A^+ + BH \rightleftharpoons AH + B^+ \qquad (4)$$

(where A^+ and B^+ are parent molecule ions), of hydride transfer[63-68]:

$$A^+ + BH \rightleftharpoons AH + B^+ \qquad (5)$$

(where A^+ and B^+ are carbonium ions) and of switching reactions[69]:

$$AB^+ + C \rightleftharpoons AC^+ + B \qquad (6)$$

In the majority of these studies, the authors are interested in obtaining, not just values for the free energy change, ΔG°, which is obtained directly from the equilibrium constant, but of reaction enthalpies, ΔH°, which can often be used to derive values for heats of formation of particular ions. In the case of proton transfer reactions, for instance, the quantity of interest is usually the "proton affinity", which is defined as the negative of the enthalpy change for the hypothetical reaction:

$$A + H^+ \rightarrow AH^+ \qquad (7)$$

The standard enthalpy change for a proton transfer reaction (2) is then just the difference in the proton affinities of the two molecules involved:

$$\Delta H^\circ \text{ (Reaction 2)} = PA(A) - PA(B) \qquad \text{II}$$

Thus, a series of determinations of relative proton affinities yields a scale of proton affinity values which can be translated to absolute proton affinities if the heats of formation of just one ion and its corresponding conjugate base are available. Several recent papers[11,14,16,19,42,47] have reported work toward establishing extensive scales of proton affinities from gas phase equilibrium constant determinations.

In the case of charge transfer reactions, the enthalpy of reaction (4) corresponds to the difference in the ionization potentials of the molecules:

$$\Delta H^{\circ} = IP(B) - IP(A) \qquad\qquad (III)$$

Thus, if values of ΔH° can be derived from the experimentally determined values of ΔG° for charge transfer, a scale of relative ionization potentials can be derived. These relative values can be related to absolute values of the ionization potentials if one or more molecules in the scale has a well-established ionization potential.

In order to derive accurate values of ΔH° from the experimentally determined values of ΔG° for a particular equilibrium, it is necessary to evaluate the entropy change for the reaction (equation I). In about one-fourth of the studies on ion-molecule equilibria which have appeared to date, equilibrium constants have been measured as a function of temperature[6-16,53-55,63-68] thus providing experimental determinations of ΔS° (or in some cases, experimental attempts at justification of the assumption that the entropy change was zero). In other studies, values of ΔS° have been estimated. One method of estimation uses standard entropy data from the literature for the neutral molecules and estimates the standard entropy of the ions by assuming that their entropies will be the same as those of the corresponding isoelectronic molecules. Another approach is to calculate entropy changes for the reaction from the ratios of the partition functions of products and reactants. In the case of proton transfer reactions, the assumption is usually made that the only significant contribution to the entropy change will arise from the ratios of rotational partition functions:

$$\Delta S_{Rot} = R\ln\frac{q_{BH}+q_A}{q_{AH}+q_B} = R\ln\left[\frac{I_{BH}+I_A}{I_{AB}+I_B}\right]^{1/2}\frac{\sigma_{AH}+\sigma_B}{\sigma_{BH}+\sigma_A} \qquad\qquad IV$$

(where the q's are the rotational partition functions, the I's are the moments of inertia, and the σ's are the symmetry numbers of the designated species). It is usually assumed that the ratio of the moments of inertia is close to unity, so that the entropy change can be estimated from an examination of the symmetry numbers alone:

$$\Delta S_{Rot} \approx R\ln \frac{\sigma_{AH} + \sigma_B}{\sigma_{BH} + \sigma_A} \qquad\qquad V$$

It has recently been pointed out[55,62] that an estimation of the entropy change for ion-molecule equilibria must, for greatest accuracy, take into account the contribution to the entropy change of the intermolecular ion-molecule interactions. The suggestion was made that this contribution to the entropy change can most easily be estimated from the expression:

$$\Delta S_{Intermolecular} = R\ln Z_f/Z_r \qquad\qquad VI$$

where Z_f and Z_r are the forward and reverse ion-molecule collision rates (usually calculated approximately by the Langevin-Gioumousis-Stevenson[70] or A.D.O. formulations.[71])

In the discussion which follows, we shall first discuss the factors which enter into the entropy change for ion-molecule equilibria, with particular attention to deriving and justifying the existence of $\Delta S_{Intermolecular}$ (equation VI). This discussion will include recent experimental results from this laboratory[16,55] which illustrate that entropy changes can be predicted from the ratios of rotational, vibrational, and electronic ("internal") partition functions added to the "intermolecular" contribution:

$$\Delta S = \Delta S_{Rot} + \Delta S_{Vib} + \Delta S_{Elec} + \Delta S_{Intermolecular} \qquad\qquad VII$$

We shall then examine some of the results in the literature with special attention to the entropy changes which were determined or assumed.

Entropy Changes in Ion-Molecule Equilibria

The entropy change for an ion-molecule equilibrium can be estimated from the ratios of the partition functions expressing the distributions of rotational, vibrational, and electronic energy in the separated reactants, plus the intermolecular contribution given by equation VI:

$$\Delta S = Rln\frac{(Q_C+) \, (Q_D)}{(Q_A+) \, (Q_B)} + \Delta S_{Intermolecular} \qquad\qquad VIII$$

where $Q_X = (q_{rot}q_{vib}q_{elec})_X$.

Equation VI can be derived in a straightforward way from kinetic theory. That is, the rate constant of a bimolecular ion-molecule reaction can be represented as

$$k_{Rn} = ZPe^{-E/RT} \qquad\qquad IX$$

where Z is the ion-molecule collision rate , P is the probability that a collision will be reactive when the exponential term approaches unity, and E is an energy barrier. Since the equilibrium constant for a reaction proceeding in both directions is equal to the ratio of the forward and reverse rate constants, one can write:

$$K_{eq} = \frac{k_f}{k_r} = \frac{Z_f P_f e^{-E_f/RT}}{Z_r P_r e^{-E_r/RT}} \qquad\qquad X$$

Then since it is reasonable to assume that most exothermic ion-molecule

reactions proceed without an observable activation energy barrier, $E_f \approx$

0, and the energy barrier in the reverse direction is just the endo-

thermicity of reaction, $E_r \approx \Delta H_{Rn}$. That is

$$K_{eq} = \frac{Z_f P_f}{Z_r P_r} e^{-\Delta H/RT} \qquad\qquad XI$$

It can be seen from a comparison of equations XI and I that

$$\Delta S = R\ln \frac{Z_f}{Z_r} + R\ln \frac{P_f}{P_r} \qquad\qquad XII$$

where the ratio $\frac{P_f}{P_r}$ includes the contributions of the ratios of rotational,

vibrational, and electronic partition functions (the "ideal" entropy).

Equations XI and XII will only be valid if there is no unanticipated

component of the ratio $\frac{P_f}{P_r}$ which effectively cancels out the ratio $\frac{Z_f}{Z_r}$.

This would be true for instance if the ratio of probabilities for sepa-

ration of the products from the ion-molecule complex entered into the ex-

pression for K_{eq} and were just the reciprocal of $\frac{Z_f}{Z_r}$. That this is not the

case has been demonstrated by data for several equilibria in which it can

be shown that both P_f and P_r are essentially unity[62]. For instance, in the

charge transfer equilibrium

$$c\text{-}C_4H_4O^+ + CH_3C_6H_5 \rightleftarrows CH_3C_6H_5^+ + c\text{-}C_4H_4O \qquad\qquad (8)$$

it is not expected that there should be any net changes in the rotatio-

nal, vibrational or electronic partition functions between reactants and

products. The forward rate constant was measured using the double resonan-

ce technique in an ion cyclotron resonance spectrometer[62] to be

$14.3 \pm 1.4 \times 10^{-10} cm^3/molecule \cdot s$. This can be compared with an estimated collision rate for the furan ion and toluene of $16.0 \pm 2.4 \times 10^{-10}$ cm^3/molecule·s. Clearly $P_f \sim 1$. In the reverse direction, the bimolecular rate constant was determined[62] to be $1.0 \pm 0.2 \times 10^{-10}$ cm^3/molecule s. If this value is used in equation IX, with a calculated value for Z of 9.8×10^{-10} cm^3/molecule·s, and the assumption that the energy barrier is just the difference in the spectroscopically determined ionization potentials of furan and toluene, it is seen that $P_r \sim 1$. Thus, the ratio $\dfrac{P_f}{P_r}$ does not contain components which cancel the contribution to the entropy change of the intermolecular interactions. These results and others demonstrating this point are presented in Table 1.

Those who are used to thinking in terms of the concept that thermodynamics treats only the properties of separated reactants and products, and that the mode of approaching equilibrium does not matter, may have some difficulty with the introduction of collision rates into thermodynamic calculations. However, it should be pointed out that the prediction of the thermochemistry of a reaction from the thermochemical properties of the separated reactants is strictly valid only when ideal gas conditions prevail, that is, only when there are no intermolecular interactions. Ordinarily the assumption is made that at low pressures intermolecular interactions are of negligible importance. Ion-molecule reactions are almost always studied at pressures which are low enough that Van der Waals interactions are negligible, i.e. in this pressure region, kineticists generally have made the valid assumption that most gases are approximately "ideal". This assumption has carried over into treatments of ion-molecule equilibria in spite of the fact that the kinetics of these reactions are described with a considerable degree of success in terms of collision

Table 1. Rates of Forward and Reverse Charge Transfer Reactions.[a]

Verification of $k_r = Z_r e^{-\Delta H/RT}$ for Selected Reactions.

Reaction	K_{eq}	k_f	Z_f	k_r	Z_r	$\Delta H =$ $RT\ln(k_r/Z_r)$	ΔH Lit[b]
		cm³/molecule·s x 10^{10}				kcal/mole	

Reaction	K_{eq}	k_f	Z_f	k_r	Z_r	$RT\ln(k_r/Z_r)$	Lit[b]
c-$C_4H_4O^+$+$C_6H_5CH_3$ ⇌ $C_6H_5CH_3^+$+c-C_4H_4O	12.6	12.8	13.2	1.0	9.8	-1.4	-1.4
$C_6H_5CH_3^+$+$C_6H_5C_2H_5$ ⇌ $C_6H_5C_2H_5^+$+$C_6H_5CH_3$	5.3	12.8	12.6	2.0	11.8	-1.1	-1.1
$C_6H_5F^+$+p-$C_6H_4F_2$ ⇌ p-$C_6H_4F_2^+$+C_6H_5F	3.5	10.2	10.4	2.4	13.6	-1.0	-1.1

[a] Reference 62. [b] Reference 74.

theories which are based on treatments of intermolecular ion-molecule interactions.

Conceptually, the introduction of collision rates into considerations of the thermodynamics of systems in which intermolecular interactions are important is perhaps best understood if we write equation X in the form:

$$(K_{eq})_{"Ideal"} = \frac{k_f/Z_f}{k_r/Z_r} = \frac{Q'_C \, Q'_D}{Q'_A \, Q'_B} \, e^{-\Delta H/RT} \qquad \text{XIII}$$

(where the Q''s are the respective partition functions expressing distributions of internal energy only). Since k/Z is the probability that a collision will be reactive ("reaction efficiency"), it can be seen that it is the ratio of these probabilities rather than the ratio of forward and reverse rate constants, k_f/k_r, which can be related to the "ideal" internal partition functions. This equation is of course equally valid for ideal gases if the ratios of collision rates are identified with the ratios of the translational partition functions. In considering chemistry involving uncharged systems, one does not usually think in these terms because of the difficulty in defining exactly what is meant by a "collision".

It should be pointed out that the existence of $\Delta S_{Intermolecular}$ (equations VI, VII, and VIII) can be derived from statistical mechanical reasoning alone, recognizing that the rotational partition function of the molecule must contain terms to account for the modification of the rotational energy of the molecule due to the approach of the ion.

Eyring, Hirshfelder, and Taylor[72] formulated the rate constant for the ion-molecule reaction

$$H_2^+ + H_2 \rightarrow H_3^+ + H \tag{9}$$

according to transition state theory making the assumption that

$$k_{Rn} = \tau \frac{kT}{h} \frac{q_{Tr}^+ q_{Rot}^+ q_{Vib}^+ q_{Elec}^+}{(q_{Tr} q_{Vib} q_{Rot} q_{Elec})_A \ (q_{Tr} q_{Vib} q_{Rot} q_{Elec})_B} \sum_{J=0}^{\infty} (2J + 1) \ e^{-\Delta E_{r_+}/kT}$$

$$\text{XIV}$$

where it is assumed that there is no activation energy, and the summation containing the exponential term represents a modification to the rotational partition function of the molecule due to the ion-molecule interaction. It can be shown[72,73] that if the vibrational frequencies and the angular momenta of the reactants do not change appreciable in the activated complex:

$$Z_{A^+B} = \frac{kT}{h} \frac{q_{Tr}^+}{(q_{Tr})_{A^+} (q_{Tr})_B} \sum_{J=0}^{\infty} (2J + 1) \ e^{-\Delta E_{r_+}/kT} \qquad \text{XV}$$

so equations VI and XIII are consistent with a strict transition state formulation of the ion-molecule rate constant.

Equation XIII can also be derived through another line of reasoning. Since the equilibrium constant as measured for the reaction:

$$A^+ + B \rightleftharpoons C^+ + D \tag{10}$$

is the observed ratio of the ion abundances (I_A^+ and I_C^+) multiplied by the ratio of the pressures or concentrations of the neutral reactant compounds (C_B and C_D):

$$K_{eq} = \frac{I_C^+ \ C_D}{I_A^+ \ C_B} \qquad \text{XVI}$$

it can easily be seen that unless ideal gas conditions prevail, or unless $k_f = k_r$, this ratio of pressures will not represent the "effec -

tive" ratio of concentrations of the two gases with respect to the reactant ions. Expressed in another way, since ion-molecule collision rates are largely determined by interactions between the charge on the ion and the various permanent or induced moments of the molecule, these collision rates are generally much greater than the analogous collision rates between uncharged species, and hence, the effective pressure or concentration of a reactant compound vis-a-vis an ion is greater than it would be if ideal gas conditions (i.e. conditions in which there are effectively no intermolecular interactions) were operative. One way to adjust the pressure ratio $\frac{C_D}{C_B}$ in equation XVI to account of this nonideality is simply to multiply the pressure of each compound by $\frac{Z_i}{Z_n}$, where Z_i is the appropriate ion-molecule collision rate, and Z_n is the corresponding collision rate for uncharged species. This results in equation XIII since the ratio $(Z_n)_{AB}/(Z_n)_{CD}$ cancels the ratios of the translational partition functions which would ordinarily appear on the right side of the equation.

Experimental Verification of Equation VII

The validity of equation VII for ion-molecule equilibria can be demonstrated by experimental determinations of the entropy change for such equilibria. Equilibrium constants for a number of charge transfer equilibria[55] (reaction 4) and proton transfer equilibria[16] (reaction 2) have been determined as a function of temperature in the range 300-400K in a pulsed ion cyclotron resonance spectrometer. For these experiments examples were chosen for which either (a) the ΔH^0_{Rn} is well established by independent means, or (b) values of ΔG^0 determined at higher temperatures are available so that the limited temperature range of the ICR experiments can be extended. In this way entropy changes could be determined more accurately.

The equilibrium constants for the charge transfer equilibrium:

$$NO^+ + C_6H_6 \rightleftharpoons C_6H_6^+ + NO \tag{11}$$

determined at temperatures from 320 to 400 K were in the range 125 to 160. Figure 1 shows a plot of ΔG determined in these experiments as a function of temperature (equation I). (Such a plot, in which the inter- cept is the enthalpy change at 300-400K for the reaction and the slope is the negative of the entropy change, is a graphical representation of the usual calculation of thermodynamic quantities from equation 1 assu- ming linearity over a limited temperature range). Since the ionization potentials of both NO and benzene have been determined spectroscopically from the limits of Rydberg series[74] (Table 2), the enthalpy change for reaction 2 at 300°K is well established as 0.40 ± 0.06 kcal/mole. If this point is included as the intercept in the Figure, it is seen that the overall slope corresponds to that which would be predicted from the ex- perimental ΔG values alone. This slope corresponds to an entropy change for reaction 11 of $+8.8 \pm 0.4$ cal/K mole. This value must be compared with the value predicted from equation VII.

The rotational entropy change is calculated from equation IV or V. In the case of C_6H_6 and $C_6H_6^+$, the Jahn-Teller distortion in the ion[75] results in a change from D_{6h} symmetry in the neutral molecule to D_{2h} symmetry in the ion, corresponding to symmetry numbers of 12 and 4. Thus, the ratio of the symmetry numbers leads to an estimate of $+2.2$ cal/K mole for ΔS_{Rot}; a more exact calculation from equation IV leads to a value of $+2.4$ cal/K mole.

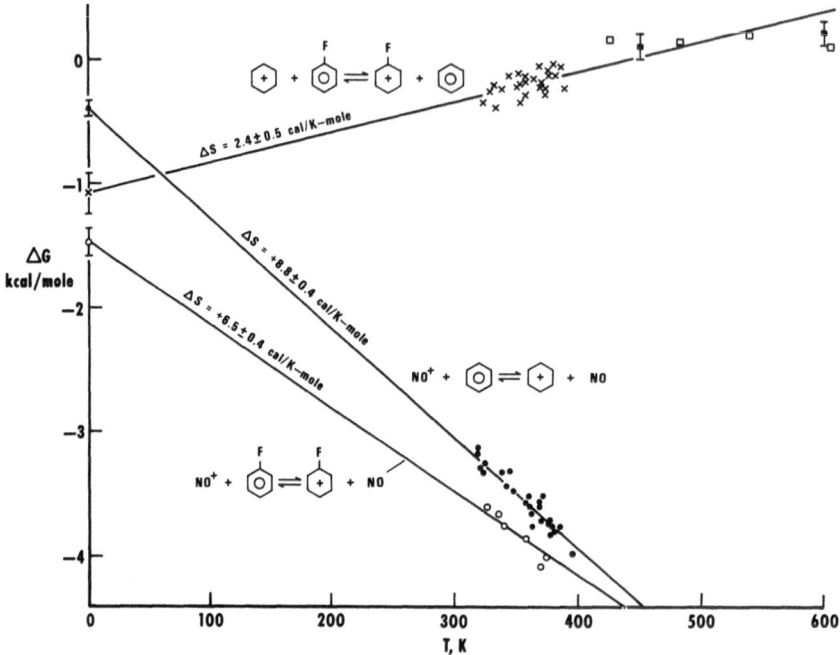

Figure 1. Values of ΔG° determined by ICR[55] as a function of the absolute temperature for the equilibria: ($NO^+ + C_6H_5F \rightleftharpoons C_6H_5F^+ + NO$) (o); ($NO^+ + C_6H_6 \rightleftharpoons C_6H_6^+ + NO$) (●); and ($C_6H_6^+ + C_6H_5F \rightleftharpoons C_6H_5F^+ + C_6H_6$) (x). For the benzene-fluorobenzene equilibrium, values from reference 54, Table 1 (■) and reference 54, Figure 1 (□) are included. Intercepts are the differences in the spectroscopic ionization energies of the compounds (Table 2).

Table 2. Ionization Energies[a] of Compounds

Compound	Ionization Energy[a]		Measured
	eV	kcal/mole	by
NO	9.263±0.0006	213.63±0.01	Spectroscopy
C_6H_6	9.247±0.002	213.23±0.05	Spectroscopy
C_6H_5F	9.200±0.005	212.1±0.1	Spectroscopy
1,2,4,5-$C_6H_2F_4$	9.39 9.350±0.004	216 215.6±0.1	Photoionization ICR[b]
m-$C_6H_4F_2$	9.34 9.332.±0.004	215.4 215.2±0.1	Calculated ICR[b]
c-C_4H_4O	8.883+0.001	201.84+0.02	Spectroscopy
$CH_3C_6H_5$	8.82±0.01	203.4+0.2	Spectroscopy, Photoionization

[a]Reference 74.

[b]Reference 55.

The vibrational entropy change is predicted from the expression:

$$\Delta S_{Vib} = R\ln \frac{\prod\limits_{i=1}^{f}(1 - e^{-h\nu_i/kT})_C^{-1}\ \prod\limits_{i=1}^{f}(1 - e^{-h\nu_i/kT})_D^{-1}}{\prod\limits_{i=1}^{f}(1 - e^{-h\nu_i/kT})_A^{-1}\ \prod\limits_{i=1}^{f}(1 - e^{-h\nu_i/kT})_B^{-1}} \qquad XVII$$

where the ν_i are the various fundamental vibrational frequencies. In the case of the NO-benzene charge transfer equilibrium (unlike other charge transfer equilibria discussed here) sufficient data[75c,76] are available to allow a calculation of ΔS_{Vib}. The value is -0.05 cal/K mole.

The contribution to the total entropy change from the ratios of electronic partition functions arises from changes in the degeneracies of the electronic states of products and reactants:

$$\Delta S_{Electronic} = R\ln \frac{g_C g_D}{g_A g_B} \qquad XVIII$$

The ground state of NO ($^2\Pi$) has a degeneracy of 4;[76] that of NO^+ ($^1\Sigma^+$) has a degeneracy of 1.[76] The ground state of benzene ($^1A_{1g}$) has a degeneracy of 1, while that of the benzene ion is expected to have a degeneracy of 2.[75] Thus, $\Delta S_{Elec} = R\ln \frac{1 \cdot 1}{2 \cdot 4} = +4.2$ cal/K mole.

The values calculated for the various contributions to the total entropy change are summarized in Table 3. The total calculated entropy change for equilibrium 11 is +8.4 cal/K mole, in good agreement with the experimentally determined value of +8.8 cal/K mole.

Figure 1 also shows the values of ΔG measured as a function of temperature for the charge transfer equilibrium:

$$C_6H_6^+ + C_6H_5F \rightleftharpoons C_6H_5F^+ + C_6H_6 \qquad (12)$$

Table 3. Comparison of Calculated and Experimentally Determined Entropy Changes for
Ion-Molecule Equilibria Shown in Figures 1-4.

$$A^+ + B \rightleftharpoons B^+ + A$$

A	B	ΔS_{Rot}^b	+	ΔS_{Elec}^c	+	ΔS_{Vib}^d	+	ΔS_{Int}^e	=	ΔS_{Total}	ΔS_{Exp}
NO	C_6H_6	+2.4		+4.2		-0.05		+1.8	=	+8.4	+8.8±0.4
C_6H_6	C_6H_5F	-2.4		0		a		+0.5	=	-1.9	-2.4±0.5
NO	C_6H_5F	0		+4.2		a		+2.3	=	+6.5	+6.5±0.4
$1,2,4,5-C_6H_2F_4$	NO	0		-4.2		a		-1.8	=	-6.0	-5.3±0.5
$m-C_6H_4F_2$	NO	0		-4.2		a		-2.3	=	-6.5	-7.1±0.5
$1,2,4,5-C_6H_2F_4$	$m-C_6H_4F_2$	0		0		a		+0.5	=	+0.5	+0.7±0.2
$c-C_6H_4O$	$CH_3C_6H_5$	0		0		a		+0.5	=	+0.5	+0.5±0.1

$$AH^+ + B \rightleftharpoons BH^+ + A$$

A	B	ΔS_{Rot}^b	+	ΔS_{Elec}^c	+	ΔS_{Vib}^d	+	ΔS_{Int}^e	=	ΔS_{Total}	ΔS_{Exp}
C_6H_5F	C_6H_6	+3.5		0		a		-0.5	=	+3.0	+2.7±0.3
C_6H_5Cl	C_6H_6	+3.5		0		a		-0.7	=	+2.8	+2.2±0.7

[a] Insufficient information available to calculate. [b] Equation IV. [c] Equation XVIII. [d] Equation XVII.

[e] Equation VI. All the entropy values are given in cal/K mole.

Included on the Figure are the data reported for this equilibrium in a high pressure mass spectrometric study.[54] The slope of the line resulting from the data reported here is consistent with an intercept of 1.08 ± 0.16 kcal/mole, which is the difference between the spectroscopically-determined ionization potentials of benzene and fluorobenzene[74] (Table 2). The experimental data from the earlier study[54] are also consistent if wide error limits are assigned to the ΔG values.

The entropy change corresponding to this slope is -2.4 cal/K mole. For this system, Z_f is 14.8 x 10^{-10} cm^3/molecule-s and Z_r is 11.5 x 10^{-10} cm^3/molecule s, so from equation VI, ΔS_{Int} = +0.5 cal/ K mole. Assuming that the structures of C_6H_5F and $C_6H_5F^+$ are essentially the same, the only contribution to ΔS_{Rot} is that arising from the Jahn-Teller distortion in the benzene ion, as discussed above; therefore, ΔS_{Rot} = -2.4 cal/ K mole. In this system, ΔS_{Elec} is equal to zero. Insufficient information is available to calculate the vibrational partition function of the fluoro-benzene ion, but it is reasonable to assume that ΔS_{Vib} is small, as it was for the NO-C_6H_6 equilibrium. Therefore, ΔS_{Total} = (+0.5 - 2.4) cal/K mole, or -1.9 cal/K mole, in reasonably good agreement with the experimental value of -2.4 ± 0.5 cal/K mole. These results are summarized in Table 3.

In the high pressure mass spectrometric study of charge transfer equilibria in aromatic compounds[54], values of ΔG were measured at 450°-600°K. Because the variations over this temperature range were of the order of only a few tenths of a kcal/mole (where 1 kcal/mole = 4.18 kjoules/mole = 0.043 eV), it was assumed that the entropy changes for charge transfer equilibria involving aromatic molecules were zero. A scale of ionization energies derived from the results of that study showed marked differences

from the scale of adiabatic ionization potentials of these aromatic compounds. For instance, in mixtures of benzene and fluorobenzene, the results seemed to indicate that the ionization energy of fluorobenzene was
higher than that of benzene by 0.2 kcal/mole, even though photoionization
and spectroscopic data indicated that the ionization potential of benzene was higher by more than 1 kcal/mole.

Because of such discrepancies, it was inferred that in certain substituted benzenes, such as C_6H_5F, formation of the ion by charge transfer involves a vertical rather than a 0-0 transition.

The results given in Figure 1 show that the free energy change
observed is consistent with an enthalpy change equal to the differences
in the adiabatic ionization potentials of benzene and fluorobenzene; the
earlier conclusion[54] that $C_6H_5F^+$ is formed with vibrational excitation
in reaction 12 is not warranted.

Also given in Figure 1 are the values of $\Delta G°$ determined as a function of temperature for the equilibrium:

$$NO^+ + C_6H_5F \rightleftharpoons C_6H_5F^+ + NO \tag{13}$$

The slope of the line resulting from these data is consistent with the
enthalpy change of 1.47 $\pm$ 0.13 kcal/mole predicted from the differences
in the adiabatic ionization potentials[74] (Table 1). The entropy change
corresponding to this slope is +6.5 $\pm$ 0.5 cal/ K mole, in excellent
agreement with the value of 6.5 cal/K mole predicted from equations IV,
V, and VII, again assuming that ΔS_{Vib} is negligible. The considerations
going into the statistical estimate of ΔS are summarized in Table 3.

Figure 2 shows the values of ΔG° measured as a function of temperature for the equilibria:

$$m\text{-}C_6H_4F_2^+ + NO \rightleftharpoons NO^+ + m\text{-}C_6H_4F_2 \qquad (14)$$

$$1,2,4,5\text{-}C_6H_2F_4^+ + NO \rightleftharpoons NO^+ + 1,2,4,5\text{-}C_6H_2F_4 \qquad (15)$$

$$1,2,4,5\text{-}C_6H_2F_4^+ + m\text{-}C_6H_4F_2 \rightleftharpoons m\text{-}C_6H_4F_2^+ + 1,2,4,5\text{-}C_6H_2F_4 \quad (16)$$

and for reaction 8 involving furan and toluene. Entropy changes predicted
for these equilibria are summarized in Table 3. In every case the predicted entropy change is in good agreement with the experimentally measured
value.

Equilibrium constants for the proton transfer reactions:

$$C_6H_5FH^+ + C_6H_6 \rightleftharpoons C_6H_7^+ + C_6H_5F \qquad (17)$$

and
$$C_6H_5ClH^+ + C_6H_6 \rightleftharpoons C_6H_7^+ + C_6H_5Cl \qquad (18)$$

were also determined as a function of temperature[16]. Figures 3 and 4 show
plots of ΔG values obtained in these experiments as a function of temperature. Also included are the values determined at 600 K in high pressure
mass spectrometric experiments[11,14]. The entropy changes measured in the
experiments are compared in the Figures and in Table 3 with entropy changes
predicted for these equilibria from statistical mechanical considerations
(equations VII, IV (or V), VI, XVII, and XVIII). For proton transfer reactions, the important contributions to the entropy change are predicted to
be the rotational entropy change (equation IV or V) and the "intermolecular" entropy change (equation VI). It is seen that the experimentally measured entropy changes agree well with the sum of the entropy changes predicted from equations V and VI assuming that protonation occurs predomina-
tely in the position para to the fluorine or to the chlorine substituent.

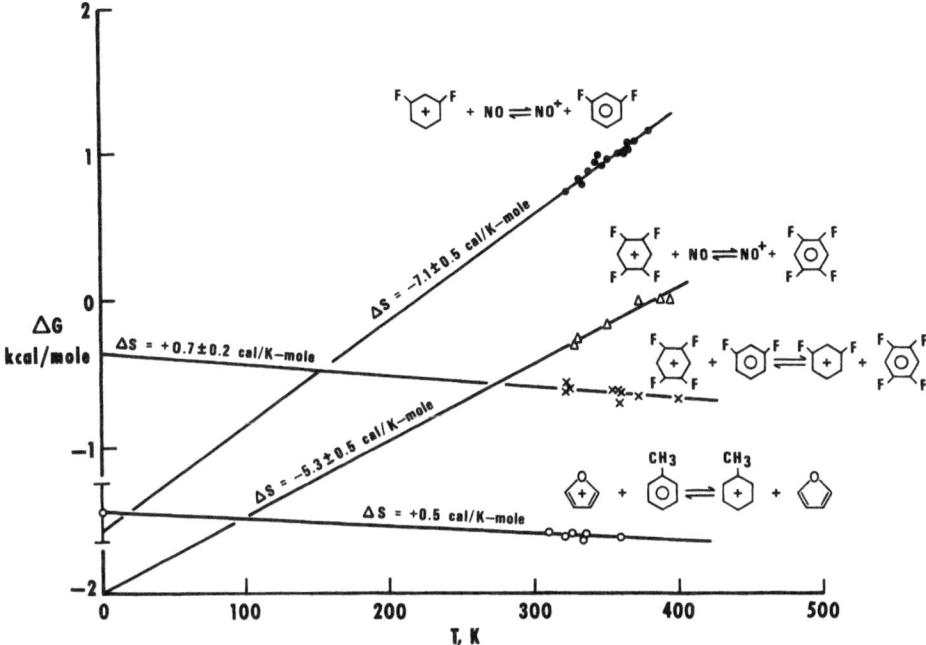

Figure 2. Values of ΔG^o as determined by ICR[55] as a function of the
absolute temperature for the equilibria: $(c\text{-}C_4H_4O^+ + CH_3C_6H_5 \rightleftarrows$
$CH_3C_6H_5^+ + c\text{-}C_4H_4O)$ (o); $(1,2,4,5\text{-}C_6H_2F_4^+ + m\text{-}C_6H_4F_2 \rightleftarrows m\text{-}C_6H_4F_2^+ +$
$1,2,4,5\text{-}C_6H_2F_4)$ (x); $(1,2,4,5\text{-}C_6H_2F_4^+ + NO \rightleftarrows NO^+ + 1,2,4,5\text{-}C_6H_2F_4)$ (Δ);
and $(m\text{-}C_6H_4F_2^+ + NO \rightleftarrows NO^+ + m\text{-}C_6H_4F_2)$ (●).

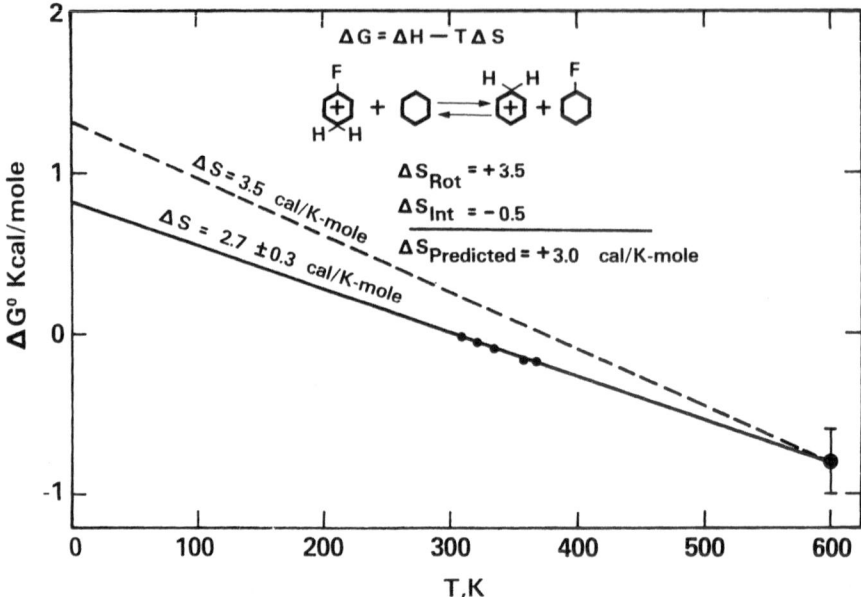

Figure 3. Values of $\Delta G°$ determined by ICR[16] as a function of the absolute temperature for the equilibrium: $(C_6H_5FH^+ + C_6H_6 \rightleftharpoons C_6H_7^+ + C_6H_5F)$. The 600 K value is from references 11 and 14. The entropy change of the reaction is the negative of the slope of the line. The dotted line indicates the entropy change for this reaction reported in reference 14. (1 kcal/mole = 4.18 kjoules).

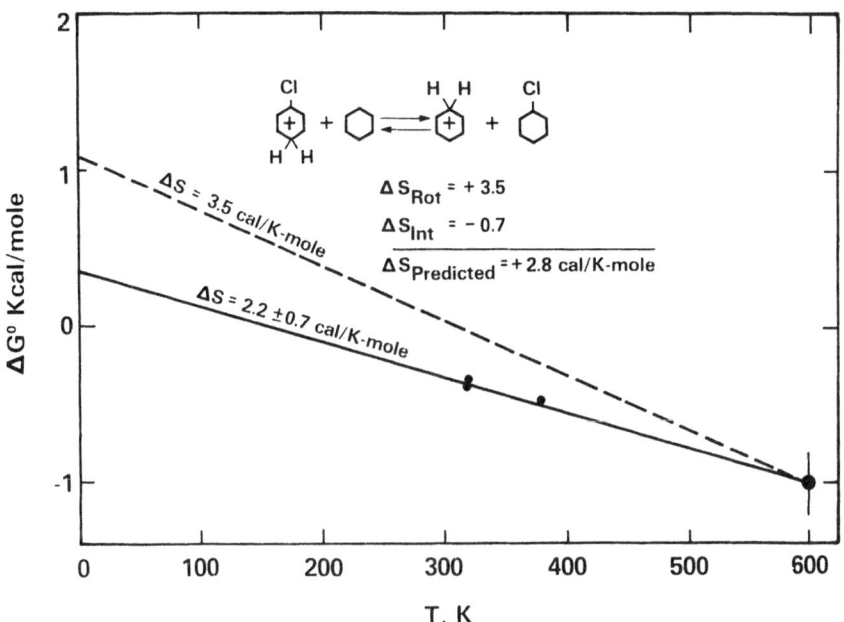

Figure 4. Values of ΔG° determined by ICR[16] as a function of the absolute temperature for the equilibrium: $(C_6H_5ClH^+ + C_6H_6 \rightleftarrows C_6H_7^+ + C_6H_5Cl)$. The 600 K value is from references 11 and 14. The entropy change of the reaction is the negative of the slope of the line. The dotted line indicates the entropy change for this reaction reported in reference 14. (1 kcal/mole = 4.18 kjoules/mole).

$$\Delta S(\text{Proton Transfer}) \approx \Delta S_{\text{Rot}} + \Delta S_{\text{Intermolecular}} \qquad \text{XIX}$$

These results are in disagreement with the entropy changes measure

for these equilibria in the high pressure mass spectrometric study,[14]

where values of + 3.5 $\pm$ 0.1 cal/K mole were reported for equilibria 14

and 15. For these equilibria involving protonated fluoro- and chloro-

benzene, the experimental entropy changes are slightly lower than those

which are predicted if one assumes that protonation occurs exclusively

in the para position. Thus, one can not rule out the formation of a small

percentage of ions protonated in the ortho position, although the data

are not of sufficient accuracy to draw any firm conclusions on the matter.

The results given in Figures 1 through 4 and the calculations summa-

rized in Table 3 demonstrate that the entropy change associated with an

ion-molecule equilibrium can be predicted rather accurately from the usual

statistical mechanical considerations if the contribution from the inter-

molecular component of the entropy change, as estimated by equation VI,

is not ignored.

It is of interest to compare the results of experimental entropy

change determinations from different laboratories. The proton transfer

equilibrium:

$$H_3O^+ + H_2S \rightleftharpoons H_3S^+ + H_2O \qquad (19)$$

has been investigated extensively using high pressure mass spectrometry[5,7,11,15],

ICR[47], and flowing afterglow[77]. In three of the high pressure mass spec-

trometric studies, the equilibrium constant was determined as a function

of temperature, so experimentally determined values of the entropy change

are available. These results show considerable discrepancy, as can be seen

by an examination of Figure 5, where all the results reported from the
different studies are displayed as conventional Van't Hoff plots. The
high pressure studies of this system carried out at relatively low temperatures (< 440 K) suffer from competing clustering reactions. In fact,
in reference 7, the investigators were not able to determine the values
of the equilibrium constant directly from the ratios of the ion currents
of H_3S^+ and H_3O^+, but had to calculate the values from a kinetic scheme
incorporating the cluster ions. The results of the other two high pressure
studies[11,15] are in approximate agreement as to the value of the equilibrium constant in the temperature range 450-650 K, but the slopes of the
plots vary considerably; the two single-temperature determinations at
~ 300 K — one from an ICR study[47], and one by flowing afterglow[77] —
are in excellent agreement with one another. It is possible to connect
these points with those reported in reference 11, thereby obtaining
ΔS^0 = -2.1 cal/K mole and ΔH^0 = -4.3 kcal/mole; a derivation based on the
three highest temperature points from reference 15 (the three lower temperature points cannot be connected to the flowing afterglow and ICR
points by a single line) leads to ΔS^0 = +1.1 cal/K mole, ΔH^0 =
-3.3 kcal/mole. Thus, at the present time, the best estimate from available results is the average of these values or ΔS = - 0.5 ± 1.6 cal/K mole,
ΔH^0 = -3.8 ± 0.5 kcal/mole.

For the proton transfer equilibrium 19, it is expected that both
ions should have the same symmetry (as do the two molecules), so that
ΔS_{Rot} should be zero. Since it is not expected that there should be any
contribution from ΔS_{Vib}, and ΔS_{Elec} is zero, the only contribution to
the entropy change should be $\Delta S_{Intermolecular}$, which is predicted from

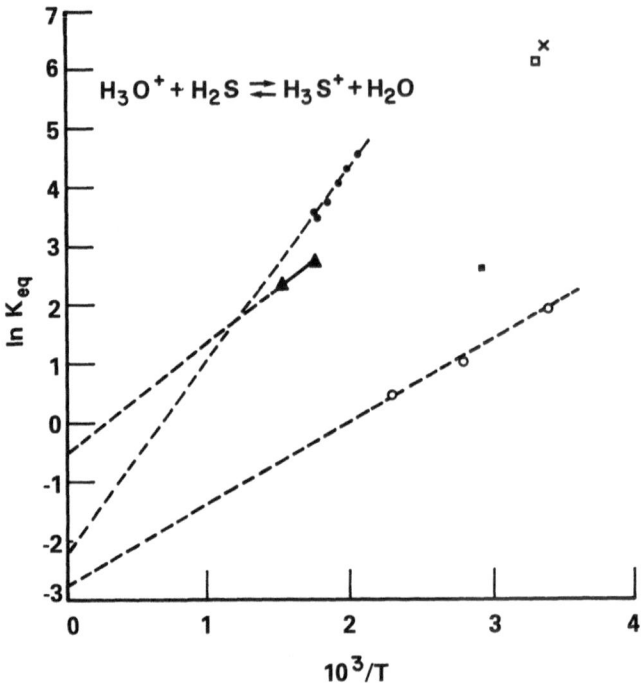

Figure 5. The logarithms of experimental equilibrium constants for the equilibrium ($H_3O^+ + H_2S \rightleftharpoons H_3S^+ + H_2O$) as a function of the reciprocal of the absolute temperature. Data are from: Hopkins and Bone (reference 7), high pressure mass spectrometry (o); Chong and Franklin (reference 5), high pressure mass spectrometry (■); Tanaka et al (reference 77), flowing afterglow (x); Wolf et al. (reference 47), ICR (□); Yamdagni and Kebarle (reference 11), high pressure mass spectrometry (▲) (Only the highest and lowest temperature results from reference 11 are explicitly presented. The intervening 12 experimental points are represented as a straight line); Meot-Ner and Field (reference 15), high pressure mass spectrometry (●) (Note that the intercept corresponds to an entropy change of -4.4 cal/K mole, rather than -2.2 cal/K mole, as reported in Table 1 of reference 15).

equation VI to be equal to -0.41 cal/K mole. This is in good agreement

with the values of - 0.5 cal/K mole derived from the results from referen-

ces 11, 15, 47, and 77. As pointed out by the authors of reference 11, it

is not possible to explain the more negative values of the entropy changes

reported in the earlier studies without postulating structures for the

H_3O^+ ion which are inconsistent with what is known about the geometry of

this ion.

Another reaction which has been extensively studied by high pressure

mass spectrometry[13,15,7,27], flowing afterglow[8], and ICR[38] is the proton

transfer equilibrium:

$$CO_2H^+ + CH_4 \rightleftharpoons CH_5^+ + CO_2 \qquad (20)$$

The results of these various studies are displayed as Van't Hoff plots

in Figure 6. The Van't Hoff plot reported in the flowing afterglow stu-

dy[8] and that derived in a high pressure mass spectrometric study[13] are

in excellent agreement. The latter study[13] was carried out to evaluate

the effect on observed equilibrium constants of applied extraction fields

in single source non-pulsed high pressure mass spectrometers; the plot

shown results from an extrapolation of experimental results to zero field

strength. These studies respectively indicate a value for ΔS° of +1.4

± 0.3 or +1.7 ± 0.3 cal/K mole, and a value for ΔH° of - 1.48 ± 0.09 or

-1.58 ± 0.06 kcal/mole. Equilibrium constants measured in the temperature

range 400-550 K in a pulsed high pressure mass spectrometer[15] are in fair-

ly good agreement with values of K_{eq} in this temperature range from the

other two studies, although the data show considerable scatter, and the

apparent Van't Hoff plot derived from these results alone indicates a

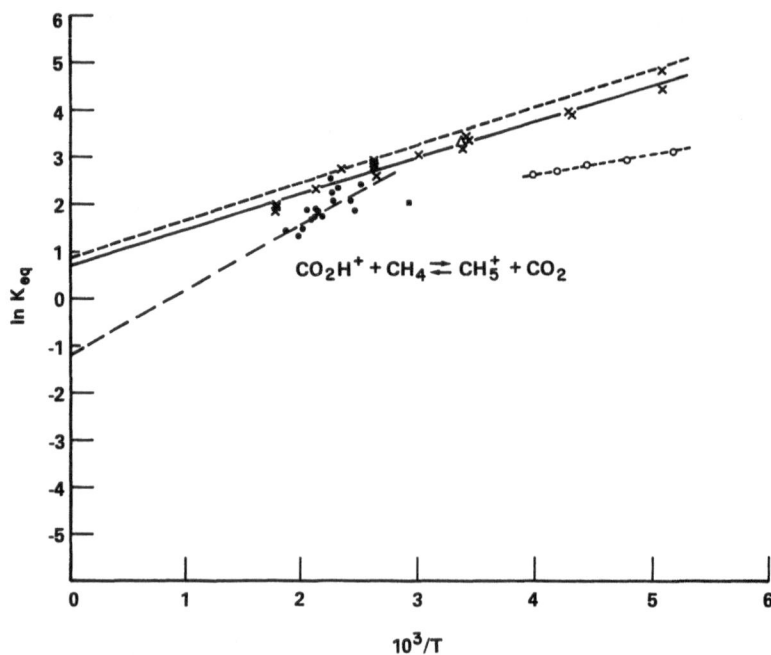

Figure 6. The logarithms of experimental equilibrium constants for the equilibrium ($CO_2H^+ + CH_4 \rightleftharpoons CH_5^+ + CO_2$) as a function of the reciprocal of the absolute temperature. Data are from: Hemsworth et al. (reference 8), flowing afterglow (x); Meot-Ner and Field (reference 15), high pressure mass spectrometry ($\bullet$); Kasper and Franklin (reference 17), high pressure mass spectrometry ($\blacksquare$); and Meisels et al. (reference 13), high pressure mass spectrometry, experimental points at lowest E/P (o) and Van't Hoff plot derived from extrapolation to E/P = 0 (----); Staley and Beauchamp (reference 38), ICR ($\triangle$).

value for ΔS of -2.4 cal/K mole. It would be difficult to explain a negative entropy change for this reaction; however, the error limits are
apparently large, since the authors[15] describe their results as being
"in reasonable agreement" with those of the flowing afterglow study[8].

The single temperature point reported in the ICR study[38] is in excellent agreement with the results of the other studies.

The predicted entropy change for reaction 20 depends on the structure which is assumed for the ions. If, following Hemsworth et al.[8], we
select the equilibrium geometry calculated by Dyczmons et al.[78] for CH_5^+,
it is predicted from equation VII that ΔS should be +2.0 to +2.2 cal/K
mole depending on whether CO_2H^+ is taken as linear or bent, respective-
ly. In this case, $\Delta S \approx \Delta S_{Rot}$, since $\Delta S_{Int} = +0.03$ cal/K mole. This predicted entropy change is in reasonably good agreement with the experi-
mental values of 1.4 ± 0.3 and 1.7 ± 0.3 cal/K mole. The experimental
accuracy is not really sufficiently good to draw conclusions about the
structure of CO_2H^+, but the experimental results would seem to favor a
linear structure for the ion, rather than the bent structure proposed
on the basis of these results in reference 8.

Conclusions

The detailed examination we have made of the results presented in
the Figures serves to illustrate several conclusions.

(1) For simple reactions involving only transfer of an atom (i.e. H^+,
H^- etc.) or an electron, the entropy change associated with ion-molecule
equilibria can be predicted accurately from statistical mechanical considerations (equations VII, IV, VI, XVII, and XVIII) provided the struc-

tures of the ions are known. The occurrence of Jahn-Teller distortions or changes in the degeneracies of electronic states, as well as the effects of the ion-molecule interactions, must be taken into account (Figures 1 through 4, and Table 3).

(2) The accuracy of experimentally-determined entropy changes of ion-molecule equilibria is generally rather low when the determinations are made over a limited temperature range (Figures 5 and 6), unless supplementary information is available (as in Figure 1 where values of ΔH° are known). The occurrence of competing reactions can be a serious complicating factor, as has been discussed recently[42,79].

For such small entropy changes as those discussed here, entropy changes derived from comparisons of values of ΔG° determined at 300 K in an ICR with values obtained at 600 K in a high pressure mass spectrometer are not sufficiently accurate to draw conclusions about the factors contributing to the entropy change. The conclusion from a recent study[47] that ΔS° (Proton Transfer) $\approx \Delta S^\circ_{Rot}$ was based on such a comparison. Table 4 shows some of the data compared in that paper[47], with the cited accuracy in ΔS° of ± 1.5 cal/K mole. It can be seen that the error limits are too large to draw any conclusions about the importance of $\Delta S_{Intermolecular}$. Certainly the existence of this contribution is consistent with the values of ΔS° deduced, and could not be precluded on the basis of these comparisons.

(3) Conclusions about the structures of ions can be based on the determination of entropy changes in ion-molecules equilibria, but the entropy change determination must be very accurate, or the differences

Table 4. Entropy Changes Derived[a] from Comparison of $\Delta G°_{300}$ (Reference 47) and $\Delta G°_{600}$ (Reference 11) for Proton Transfer: $AH^+ + B \rightleftharpoons BH^+ + A$.

| | | $\Delta G°_{300}$ | $\Delta G°_{600}$ | $\Delta S°$ | ΔS_{Rot} | ΔS_{Int} |
| | | kcal/mole | | | cal/K-mole | |
A	B	(Ref. 47)	(Ref. 11)		(Eqn. V)	(Eqn. VI)
MeCN	EtOH	0.3	0.6	-1.0±1.5	0.0	-1.1
MeCHO	EtOH	-1.7	-1.4	-1.0±1.5	0.0	-0.5
EtOH	MeOH	4.5	4.7	-0.7±0.5	0.0	-0.2
MeOAc	MeOH	13.2	13.3	-0.3±1.5	0.0	-0.3
HCO$_2$Me	MeOH	5.5	5.7	-0.7±1.5	0.0	-0.5
Et$_2$O	NH$_3$	-3.7	-2.6	-3.7±1.5	-4.1	-0.2
Me$_2$CO	NH$_3$	-7.2	-6.1	-3.7±1.5	-4.0	-1.2
EtCHO	NH$_3$	-13.5	-12.8	-2.3±1.5	-2.7	-1.1
EtOH	NH$_3$	-14.8	-13.9	-3.0±1.5	-2.7	-0.5

[a]Table extracted from Tables VII and VIII of Reference 47.

in values of $\Delta S°$ to be expected from different possible ion structures must be very large.

For example, the entropy change of +8.8 cal/K mole determined[55] for the charge transfer equilibrium in NO-benzene mixtures (reaction 11) is only consistent with a benzene ion which has undergone a Jahn-Teller distortion. If $C_6H_6^+$ had the same structure as benzene, the predicted entropy change would be +7.3 cal/K mole ($\Delta S_{Rot} = 0$, $\Delta S_{Elec} = +5.5$, $\Delta S_{Int} = 1.8$ cal/K mole) rather than +8.3 cal/K mole.

Abboud, Hehre, and Taft[53] were able to show that the $C_7H_7^+$ ion formed in benzyl chloride has the benzyl ion structure since the entropy change for the equilibrium:

$$C_7H_7^+ + (CH_3)_3CCl \rightleftharpoons (CH_3)_3C^+ + C_6H_5CH_2Cl \tag{21}$$

was determined to be +0.5 ± 1.5 cal/K mole. This is consistent with the value to be expected for reaction of a benzyl ion (+0.2 cal/K mole), but inconsistent with the value of +4.1 cal/K mole which would be expected if $C_7H_7^+$ had the tropylium structure.

In this regard it should be noted that conclusions apparently cannot be made about ion structures based on a comparison of measured entropy changes with those predicted from the changes in the standard entropies of reactants and products in these systems. In such calculations, the assumption is usually made that the standard entropy of the ionic species is equal to that of the neutral molecule which is isoelectronic to it. Even in cases, such as the proton transfer equilibrium in CO_2-CH_4 mixtures (Figure 6) where $\Delta S_{Intermolecular}$ is negligible, the use of the standard entropy data for the ionic species does not seem to be sufficiently

accurate to distinguish between alternate structures by comparison with
the experimentally determined entropy changes.

References

1. For reviews see: (a) P. Kebarle, Chapter 7 in "Ion-Molecule Reac-
tions (J. L. Franklin, ed.), Plenum Press, New York, (1972); (b)
E. E. Ferguson, Chapter 8 in "Ion-Molecule Reactions (J. L. Franklin,
ed.), Plenum Press, New York (1972); (c) P. Kebarle in "Interactions
Between Ions and Molecules (P. Ausloos, ed.), Plenum Press, New
York (1975)

2. M. T. Bowers, D. H. Aue, H. M. Webb, and R. T. McIver, Jr., J. Am.
Chem. Soc. 93, 4314 (1971)

3. R. T. McIver, and J. R. Eyler, J. Am. Chem. Soc. 93, 6334 (1971)

4. A. E. Roche, M. M. Sutton, D. K. Bohme, and H. I. Schiff, J. Chem.
Phys. 55, 5480 (1971)

5. S.-L. Chong, R. A. Myers, Jr., and J. L. Franklin, J. Chem. Phys.
56, 2427 (1971)

6. J. B. Briggs, R. Yamdagni, and P. Kebarle, J. Am. Chem. Soc. 94,
5128 (1972)

7. J. M. Hopkins and L. I. Bone, J. Chem. Phys. 58, 1473 (1973)

8. R. S. Hemsworth, H. W. Rundle, D. K. Bohme, H. I. Schiff, D. B. Dunkin,
and F. C. Fehsenfeld, J. Chem. Phys. 59, 61 (1973)

9. R. Yamdagni and P. Kebarle, J. Am. Chem. Soc. 95, 4050 (1973)

10. K. Hiroaka, R. Yamdagni, and P. Kebarle, J. Am. Chem. Soc. 95,
6833 (1973)

11. R. Yamdagni and P. Kebarle, J. Am. Chem. Soc. 98, 1320 (1976)

12. F. C. Fehsenfeld, W. Lindinger, H. I. Schiff, R. S. Hemsworth, and
D. K. Bohme, J. Chem. Phys. 64, 4887 (1976)

13. G. G. Meisels, R. K. Mitchum, and J. P. Freeman, J. Phys. Chem. 80,
2845 (1976)

14. Y. K. Lau and P. Kebarle, J. Am. Chem. Soc. 98, 7452 (1976)

15. M. Meot-Ner and F. H. Field, J. Chem. Phys. 66, 4527 (1977)

16. K. G. Hartman and S. G. Lias, Int. J. Mass Spectrom. Ion Phys.,
submitted for publication

17. S. F. Kasper and J. L. Franklin, J. Chem. Phys. 56, 1156 (1972)

18. W. G. Henderson, M. Taagepera, D. Holtz, R. T. McIver, Jr., J. L.
Beauchamp and R. W. Taft, J. Am. Chem. Soc. 94, 4728 (1972)

19. D. H. Aue, H. M. Webb, and M. T. Bowers, J. Am. Chem. Soc. 94, 4726 (1972)

20. S.-L. Chong and J. L. Franklin, J. Am. Chem. Soc. 94, 6347 (1972)

21. E. M. Arnett, F. M. Jones, III, M. Taagepera, W. G. Henderson, J. L. Beauchamp, D. Holtz, and R. W. Taft, J. Am. Chem. Soc. 94, 4724 (1972)

22. D. H. Aue, H. M. Webb, and M. T. Bowers, J. Am. Chem. Soc. 95, 2699 (1973).

23. D. K. Bohme, R. S. Hemsworth, H. W. Rundle, and H. I. Schiff, J. Chem. Phys. 58, 3504 (1973)

24. R. T. McIver, Jr., J. A. Scott, and J. M. Riveros, J. Am. Chem. Soc. 95, 2706 (1973)

25. D. K. Bohme, R. S. Hemsworth, and H. W. Rundle, J. Chem. Phys. 59, 77 (1973)

26. D. K. Bohme, R. S. Hemsworth, and H. W. Rundle, J. Chem. Phys. 59, 77 (1973)

27. A. G. Harrison and A. S. Blair, Int. J. Mass Spectrom. Ion Phys. 12, 175 (1973)

28. R. T. McIver, Jr., and J. H. Silvers, J. Am. Chem. Soc. 95, 8462 (1973)

29. P. F. Fennelly, R. S. Hemsworth, H. I. Schiff, and D. K. Bohme, J. Chem. Phys. 59, 6405 (1973)

30. R. H. Staley and J. L. Beauchamp, J. Am. Chem. Soc. 96, 1604 (1974)

31. R. Yamdagni and P. Kebarle, Can. J. Chem. 52, 861 (1974)

32. R. T. McIver, Jr., and J. Scott Miller, J. Am. Chem. Soc. 96, 4323 (1974)

33. T. B. McMahon and P. Kebarle, J. Am. Chem. Soc. 96, 5940 (1974)

34. E. M. Arnett, L. E. Small, R. T. McIver, Jr., and J. Scott Miller, J. Am. Chem. Soc. 96, 5638 (1974)

35. D. K. Bohme, G. I. Mackay, H. I. Schiff, and R. S. Hemsworth, J. Chem. Phys. 61, 2175 (1974)

36. W. J. Hehre, R. T. McIver, Jr., J. A. Pople, and P. v. R. Schleyer, J. Am. Chem. Soc. 96, 7162 (1974)

37. H. I. Schiff and D. K. Bohme, Int. J. Mass Spectrom. Ion Phys. 16, 167 (1975)

38. R. H. Staley and J. L. Beauchamp, J. Chem. Phys. 62, 1998 (1975)

39. J. D. Payzant, H. I. Schiff, and D. K. Bohme, J. Chem. Phys. 63, 149 (1975)

40. J. F. Wolf, P. G. Harch, R. W. Taft, and W. J. Hehre, J. Am. Chem. Soc. 97, 2902 (1975)

41. J. F. Wolf, P. G. Harch, and R. W. Taft, J. Am. Chem. Soc. 97, 2904 (1975)

42. D. H. Aue, H. M. Webb, and M. T. Bowers, J. Am. Chem. Soc. 98, 311 (1976)

43. G. I. Mackay, R. S. Hemsworth, and D. K. Bohme, Can. J. Chem. 54, 1624 (1976)

44. J. L. Devlin, III, J. F. Wolf, R. W. Taft, and W. J. Hehre, J. Am. Chem. Soc. 98, 1990 (1976)

45. L. F. Wolf, J. L. Devlin, III, D. J. DeFrees, R. W. Taft, and W. J. Hehre, J. Am. Chem. Soc. 98, 5097 (1977)

46. S. K. Pollack, J. F. Wolf, B. A. Levi, R. W. Taft, and W. J. Hehre, J. Am. Chem. Soc. 99, 1350 (1977)

47. J. F. Wolf, R. H. Staley, I. Koppel, M. Taagepera, R. T. McIver, Jr., J. L. Beauchamp, and R. W. Taft, J. Am. Chem. Soc. 99, 5417 (1977)

48. P. Ausloos and S. G. Lias, J. Am. Chem. Soc. 99, 4198 (1977)

49. P. Ausloos and S. G. Lias, Chem. Phys. Lett., in press

50. T. B. McMahon, R. J. Blint, D. P. Ridge, and J. L. Beauchamp, J. Am. Chem. Soc. 94, 8934 (1972)

51. R. J. Blint, T. B. McMahon, and J. L. Beauchamp, J. Am. Chem. Soc. 96, 1269 (1974)

52. J. A. Jackson, S. G. Lias and P. Ausloos, J. Am. Chem. Soc., in press

53. J.-L. M. Abboud, W. J. Hehre, and R. W. Taft, J. Am. Chem. Soc. 98, 6072 (1976)

54. M. Meot-Ner and F. H. Field, Chem. Phys. Lett. 44, 484 (1976)

55. S. G. Lias, Chem. Phys. Lett., in press

56. V. G. Aninich, M. T. Bowers, R. M. O'Malley, and K. R. Jennings, Int. J. Mass Spectrom. Ion Phys. 11, 99 (1973)

57. V. G. Anicich and M. T. Bowers, Int. J. Mass Spectrom. Ion Phys. 13, 351 (1974)

58. L. W. Sieck and R. Gorden, Jr., Int. J. Mass Spectrom. Ion Phys. 19, 269 (1976)

59. S. G. Lias, P. Ausloos, and Z. Horvath, Int. J. Chem. Kinetics, VIII 725 (1976)

60. J. F. Wolf, J. L. Devlin, III, R. W. Taft, M. Wolfsberg, and W. J. Hehre, J. Am. Chem. Soc. 98, 287 (1976)

61. T. B. McMahon, P. G. Miasek, and J. L. Beauchamp, Int. J. Mass Spectrom. Ion Phys. 21, 63 (1976)

62. S. G. Lias and P. Ausloos, J. Am. Chem. Soc. 99, 4831 (1977)

63. J. J. Solomon and F. H. Field, J. Am. Chem. Soc. 95, 4483 (1973)

64. J. J. Solomon, M. Meot-Ner, and F. H. Field, J. Am. Chem. Soc. 96, 3727 (1974)

65. J. J. Solomon snd F. H. Field, J. Am. Chem. Soc. 97, 2625 (1975)

66. M. Meot-Ner, J. J. Solomon, and F. H. Field, J. Am. Chem. Soc. 98, 1025 (1976)

67. J. J. Solomon and F. H. Field, J. Am. Chem. Soc. 98, 1567 (1976)

68. A. Goren and B. Munson, J. Phys. Chem. 80, 2848 (1976)

69. J. L. Beauchamp, M. C. Caserio, and T. B. McMahon, J. Am. Chem. Soc. 96, 6243 (1974)

70. (a) P. Langevin, Ann. Chim. Phys. 5, 245 (1905); (b) G. Gioumousis and D. P. Stevenson, J. Chem. Phys. 29, 294 (1958)

71. (a) T. Su and M. T. Bowers, J. Am. Chem. Soc. 95, 1370 (1973); (b) T. Su and M. T. Bowers, Int. J. Mass Spectrom. Ion Phys. 12, 347 (1973); (c) T. Su and M. T. Bowers, Int. J. Mass Spectrom. Ion Phys. 17, 211 (1975)

72. H. Eyring, J. O. Hirschfelder, and H. S. Taylor, J. Chem. Phys. 4, 479 (1936)

73. K. Yang and T. Ree, J. Chem. Phys. 35, 588 (1961)

74. (a) H. M. Rosenstock, K. Draxl, B. W. Steiner, and J. T. Herron, J. Phys. and Chem. Ref. Data 6, Suppl. 1(1977); (b) G. G. Hall, Faraday Disc., Chem. Soc. 54, 7 (1972)

75. (a) L. Asbrink, E. Lindholm and O. Edquist, Chem. Phys. Lett. $\underline{5}$, 609 (1970); (b) L. Salem, "The Molecular Orbital Theory of Conjugated Systems," W. A. Benjamin, Inc., New York (1966); (c) G. Herzberg, "Electronic Spectra of Polyatomic Molecules", Van Nostrand, Princeton (1966)

76. (a) G. Herzberg, "Spectra of Diatomic Molecules," Van Nostrand, New York (1950); (b) D. W. Turner, C. Baker, A. D. Baker, and C. R. Brundle, "Molecular Photoelectron Spectroscopy," Wiley-Interscience, New York (1970)

77. K. Tanaka, G. I. Mackay, and D. K. Bohme, Can J. Chem., in press.

78. V. Dyczmons, V. Staemmler, and W. Kutzelnigg, Chem. Phys. Lett. $\underline{5}$, 361 (1970).

79. W. R. Davidson, M. T. Bowers, T. Su, and D. H. Aue, Int. J. Mass Spectrom. Ion Phys. $\underline{24}$, 83 (1977)

Pulsed ICR Studies with a One-Region
Trapped Ion Analyzer Cell

Robert T. McIver, Jr.
Department of Chemistry
University of California
Irvine, California 92717

Introduction

The first mass spectrometer based upon the cyclotron resonance principle was developed in 1949.[1,2] At that time the device was called an
omegatron mass spectrometer. During the 1950's a number of groups studied
the theory of its operation in the hope of developing it as a compact,
light-weight mass spectrometer. It was successfully applied to residual
gas analysis in ultrahigh vacuum[3-5] and to airborne sampling of the upper
atmosphere.[6] But in spite of a number of desireable features, the omegatron was never widely used because of its low mass range and low mass resolution; it was limited to unit resolution of about m/e 50.

In recent years significant progress has been made in understanding
how best to apply the cyclotron resonance principle to mass spectrometry.
Ion trapping cells can now store ions efficiently for several seconds, and
the mass resolution and mass range are comparable now to the best double
focusing sector mass spectrometers. The International Symposium on Ion
Cyclotron Resonance Spectrometry in September, 1976, provided an unusually
good opportunity to summarize these developments and discuss future trends.
The purpose of this paper is to focus on recent technological advances rather than on chemical applications. Following a brief historical discussion

of how the techniques of ICR have evolved, a detailed description is given of the pulsed ICR spectrometers which have been developed at the University of California, Irvine, during the last four years. Since ICR spectrometers are not currently available commercially specific details of the experiments such as which diffusion pump oil to use, plating materials for the cells, types of detectors, etc. will be included in this paper.

Historical Perspective

All ion cyclotron resonance spectrometers are based on the classical motion of charged particles in magnetic and electric fields. The equation of motion for an ion of mass m and charge q in the presence of a magnetic field $\vec{B}$ and an electric field $\vec{E}$ is

$$F = ma$$

$$m \frac{d\vec{v}}{dt} = q \, (\vec{v} \times \vec{B}) + q\vec{E} \tag{1}$$

where $\vec{v}$ is the velocity of the ion. In order to achieve maximum mass resolution, most ICR spectrometers have utilized strong, highly homogeneous magnetic fields. Laboratory electromagnets are used for NMR and ESR experiments and are presently used in most ICR spectrometers.[7,8] An omegatron mass spectrometer was made using a permanent magnet,[6] and a solenoidal electromagnet was used by Wobschall, et al.[9-12] But both permanent magnets and conventional solenoidal electromagnets seem to suffer from low field strength (only a few thousand gauss) and inadequate homogeneity. Superconducting solenoids, on the other hand, were discussed at the Symposium as viable alternatives to laboratory electromagnets. The cost of a superconducting magnet system has decreased greatly in recent years, and significant improvements in the mass resolution, mass range, and ion storage efficiency of an ICR spectrometer can be expected at the higher field strengths which are available.

The various ICR techniques which have evolved during the last thirty years can be best categorized and distinguished by the last term in Eq. (1), the electric field $\vec{E}$. The particular type of electric field imposed upon the ions is the primary factor which determines their drift motion and resonance motion. The omegatron mass spectrometer shown in Fig. 1 utilizes

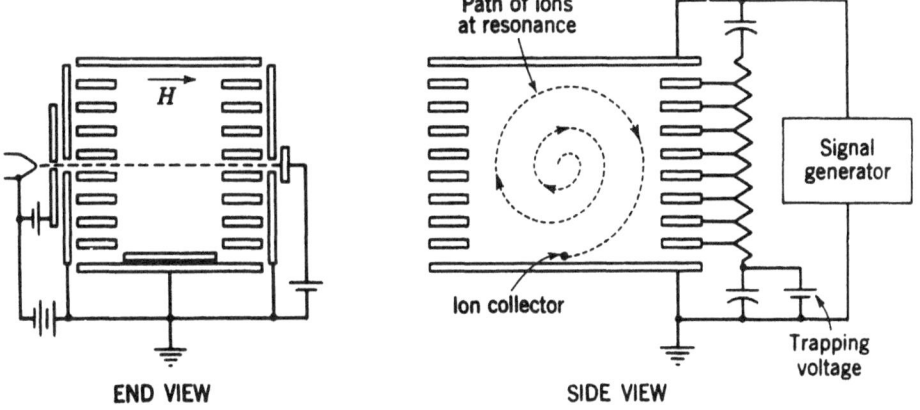

Fig. 1: Omegatron mass spectrometer developed by Sommer, Thomas, and Hipple.

a series of guard rings to obtain a linearly polarized alternating electric field perpendicular to the magnetic field.[1,2] When the cyclotron frequency of an ion, $\omega = qB/m$, is equal to the frequency of the alternating electric field, a resonance condition is established and the ion is accelerated to larger orbital radius until it finally impinges on a collector electrode. The current of resonant ions is measured with an electrometer. In addition to the alternating electric field, there is an electrostatic field which limits the motion of the ions parallel to the magnetic field. Small DC voltages on the electrodes perpendicular to the magnetic field create an electrostatic potential well which alters slightly the resonant frequency of the ions and causes them to undergo harmonic oscillatory motion in the direction parallel to the magnetic field.[2,13-15] These two types of electric fields - an alternating electric field perpendicular to the magnetic field for accelerating resonant ions and approximately quadrupolar electrostatic field parallel to the magnetic field for trapping ions - are features shared by all ICR spectrometers.

Ion-molecule collisions severely degrade the performance of an omegatron mass spectrometer because detection of resonant ions is based upon expansion of the cyclotron orbit until the ion strikes the narrow collector electrode. Collisions with the background gas shift the center of gyration and can scatter the ions parallel to the magnetic field, both effects causing them to miss the collector electrode. An ion cyclotron resonance spectrometer developed in the 1960's by Wobschall avoided this problem by using a radiofrequency (RF) bridge circuit to detect ion cyclotron resonance power absorption.[9] The apparatus, shown in Fig. 2, was used to study

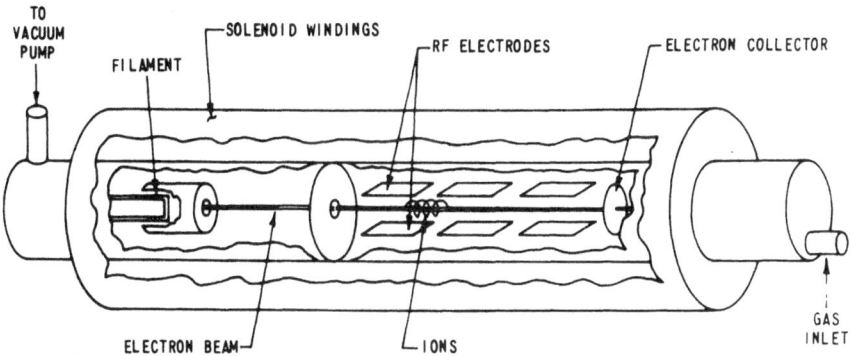

TO
VACUUM
PUMP

FILAMENT

SOLENOID WINDINGS

RF ELECTRODES

ELECTRON COLLECTOR

ELECTRON BEAM

IONS

GAS
INLET

Fig. 2. Schematic diagram of the ion cyclotron resonance apparatus used by Wobschall, Graham and Malone with a solenoid magnet.

a number of ion-molecule collision processes of atmospheric interest, such as symmetrical charge exchange and ion-atom transfer reaction.[10-12]

In 1965 Varian Associates, Inc. produced the first commercial ICR spectrometer for J. D. Baldeschwieler at Stanford University. A three section ICR cell, shown in Fig. 3, was developed to eliminate space charge oscillations and associated ion resonance shifts caused by the electron beam. A third type of electric field, a linear electrostatic field perpendicular to the magnetic field, is utilized to drift the ions from one region to the next. Application of small DC voltages V_{UPPER} and V_{LOWER} on the upper and lower plates, respectively, creates an approximately linear

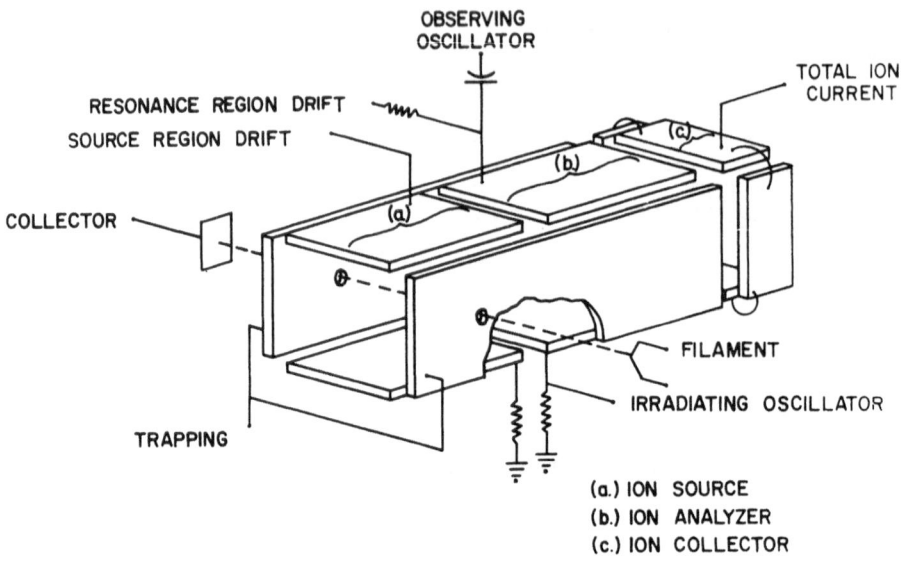

Fig. 3: A three section ICR drift cell. The magnetic field lines are parallel to the upper and lower plates of the cell.

electric field,

$$E = \frac{(V_{UPPER} - V_{LOWER})}{d}, \tag{2}$$

where d is the separation between the plates. Substituting this into Eq. (1)
and solving the equation of motion shows that a drift velocity v_d = -E/B
perpendicular to both the magnetic field and the electric field is superimposed on the cyclotron motion of the ions. As an example, for an ICR cell
with upper and lower plates separated by 0.01 m, a voltage difference of
0.5 V, and a magnetic field strength of 1.0 T, the drift velocity is 50 m
sec^{-1}. Since most three section drift cells are about 0.2 m long, an ion will

take about 3 msec to drift from the source region where it is formed by
electron impact through the analyzer region and to the collector region.
A marginal oscillator is connected to the analyzer region, and ions are
detected when their cyclotron frequency is the same as the frequency of
the marginal oscillator.[8] Even at pressures as high as 10^{-4} Torr acceptab-
le signals can be obtained with a marginal oscillator because it responds
to the power absorbed by resonant ions rather than to resonant ion current
impinging on a collector electrode, as in an omegatron. This is the main
feature which has enabled the ICR drift cell to be used in investigations
of a wide variety of ion-molecule reaction processes.[16] A four-section
drift cell has also been developed.[17-19] Experimental measurements of
ion-molecule reaction rate constants is simplified in this cell by slowly
drifting the ions through a reaction region inserted between the source
region and the analyzer region.

Spatial separation of ion production and ion mass analysis regions
eliminates the electron beam space charge effects encountered in both the
omegatron and the Wobschall apparatus. However, drift of the ions from
one region to another causes a number of experimental problems which are
most evident in quantitative studies of ion-molecule rate constants and
equilibrium constants. First, the residence time of ions in each region of
the drift cell is an essential parameter in reaction rate studies. In principle, the residence time τ can be calculated from the equation

$$\tau = \frac{\ell Bd}{(V_{UPPER} - V_{LOWER})} \tag{3}$$

where ℓ is the length of the region, and the other parameters are as defined above. However, a number of factors can cause Eq. (3) to predict inac-
curate residence times. The electron beam tends to trap positive ions around
it, thereby increasing the residence time in the source region. Also, when
polar molecules absorb on the surface of the cell plates, surface charges
can build up and alter the drift motion of the ions. Drift cells usually
require extensive "tuning" of the plate voltages to counteract the effect
of surface charges. The only reliable method for determining the residence
time is to measure it using a pulsing scheme.[20] Even with reliable values
for the residence time, however, measurement of rate constants with a drift
cell is still quite complicated. Since a mass spectrum is normally obtained at a constant marginal oscillator frequency ω_1 while scanning the magnetic field strength, Eq. (3) needs to be modified in practice by $B = \omega_1 m/q$. This gives

$$\tau_{MS} = \frac{\ell d\omega_1}{(V_{UPPER} - V_{LOWER})} \frac{m}{q} .$$

As the magnetic field is scanned, all the terms in brackets are constant
and the residence time is proportional to m/q. Thus, high mass ions have
a longer residence time and a greater opportunity to react than do low
mass ions. This feature of the drift cell experiment also causes excessive
space charge effects to occur at high magnetic field strength.

A second problem with using a drift cell for quantitative rate measurements is that the marginal oscillator signal for a reactive ion is an
average power absorption for a distribution of ions along the length of
the analyzer region. For a simple $A^+ \rightarrow B^+$ process, the number of A^+ ions

leaving the source region and entering the analyzer region is greater than the number of A^+ ions leaving the analyzer region, whereas the opposite is true for the product ions B^+. This situation leads to complex integrals even for simple ion-molecule reaction schemes.[21-23]

A third problem is that while the ions are reacting in the analyzer region their kinetic energy can be significantly perturbed by the marginal oscillator; the ions are detected while they are reacting. This is an undersirable situation since measurement of thermal rate constants is usually of greatest interest. High pressures or perhaps addition of a nonreactive buffer gas help to damp the excess ion kinetic energy. This problem is also minimized by using a four-section cell with a long reaction region.[17-19]

All of the problems discussed above were important in stimulating the development of new ICR cell designs more suitable for quantitative kinetic and equilibrium studies. In 1970 a trapped ion analyzer cell having only one spatial region and a pulsed mode of operation was introduced.[24] It soon became apparent that the full range of ICR experiments could be performed with a one-region cell by dispersing events in "time" rather than in "space."[25] The key feature which makes this a viable approach is the ability to create electric fields which are effective in trapping ions for times on the order of seconds. In terms of Eq. (1), electrostatic fields $\vec{E}$ must be created which function in conjunction with the magnetic field to trap the ions in three dimensions. One method of accomplishing this is the one region trapped ion analyzer cell shown in Fig. 4. Gaseous positive or negative ions are produced in the center of the cell by an electron beam which is accelerated from the filament, through the cell,

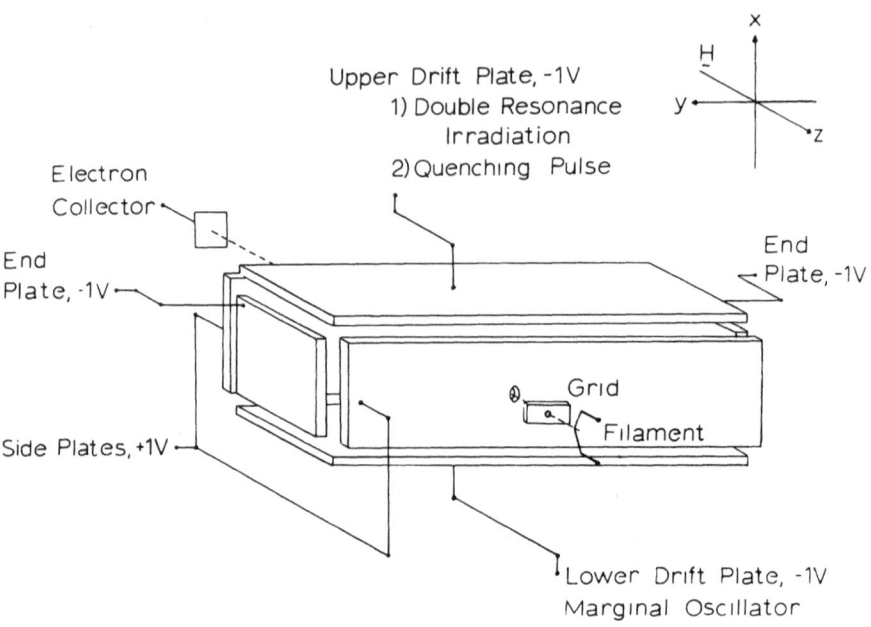

Fig. 4: A one-region trapped ICR cell shown with DC voltages suitable for storing positive ions.

and to the electron collector plate. A strong homogeneous magnetic field collimates the electron beam and directs it in the direction parallel to the magnetic field lines. The voltage applied to the repeller grid can be pulsed to gate the electron beam on and off. In the direction perpendicular to the magnetic field the motion of the ions is constrained to circular cyclotron orbits. In the direction parallel to the magnetic field, the ions are trapped by approximately harmonic oscillation motion in an electrostatic potential well. For positive ions a potential well of this type is produced by a positive voltage on the two side plates and a negative voltage on the upper plate, lower plate, and the two end plates.

The end plates are an essential feature because they cause the ions to slowly drift back and forth from one end of the cell to the other.[26] An alternating electric field perpendicular to the magnetic field is created by applying an RF voltage to the lower plate. When the frequency of the RF electric field is equal to the cyclotron frequency of an ion, a resonance condition is established and the ion is accelerated to higher kinetic energy. However, when the frequency of the RF electric field is not equal to the cyclotron frequency, the kinetic energy of the ions is not appreciably perturbed. Pulsed ICR techniques have been developed whereby ions are generated by electron impact, stored in a static magnetic ion trap such as the one shown in Fig. 4 for a specified reaction period (usually several tenths of a second), detected by pulsed RF irradiation at the cyclotron frequency of the ions, and then quenched at the walls of the trap by inverting the polarity of the electrostatic trapping well.[25]

In 1972 a versatile ICR cell which could be operated as a drift cell or as a trapped ion cell was introduced.[27] In design this cell is quite similar to the three-section drift cell, Fig. 3. For trapped ion studies is it necessary to change some of the wiring to the cell plates, add an end plate to prevent ions from drifting out of the source region in the direction opposite from the analyzer region, and insert fine wire mesh grids over the entrance and exit holes for the electron beam to screen out penetrating electric fields from the collector plate and the filament. The electronics of the spectrometer are also modified to provide for pulsing the voltages applied to the various plates of the cell.

Ion trapping cells have greatly simplified the measurement of reaction rate constants since the reaction time for the ions is controlled by

electronic timing circuits rather than the drift velocity through the cell.
Furthermore, storage times of several seconds (rather than a few msec in
a drift cell) allow for the study of equilibrium acidities and basicities
of molecules in the gas phase,[28-31] and the rate constants for "slow" ion-
molecule reactions.[32,33] A detailed description of the one-region trapped
ion cells developed in our laboratory is given below, and some examples
are given to illustrate how the experiments are done.

Description of a One-Region Trapped ICR Cell

A block diagram of the pulsed ICR spectrometer constructed in our laboratory at the University of California, Irvine, is shown in Fig. 5. There

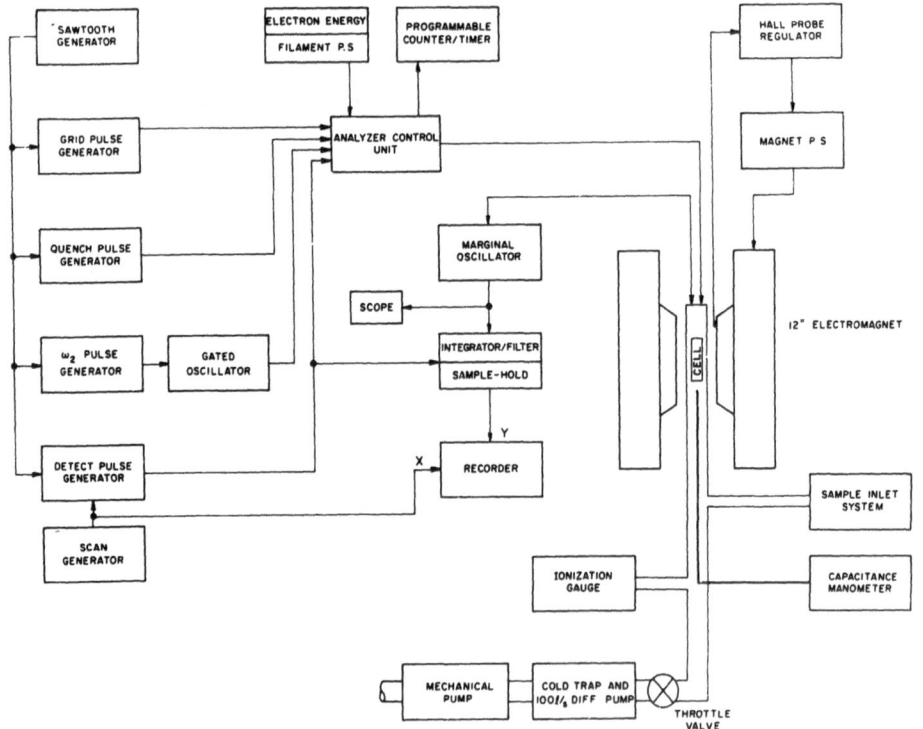

Fig. 5. Block diagram of a pulsed ICR spectrometer developed by the author
at the University of California, Irvine.

are basically three main parts - a magnet, a vacuum system, and an electronics console. A Varian Associates V-3601 low impedance 30 cm diameter electromagnet with a 4.5 cm gap and a V-2503 regulated power supply are used
in conjunction with a Fieldial I controller and Hall probe sensor to scan
the magnetic field strength up to 18kG. An electromagnet is required presently because detection of the ions is accomplished with a fixed frequency, marginal oscillator circuit. In the future, however, it may be
more practical to use a high field permanent magnet and a variable frequency circuit for detecting the cyclotron resonance of the ions.[34-37]

The vacuum system of the spectrometer was designed to provide for
accurate determination of the pressure in the analyzer cell, a high pumping capacity for rapid pump-down between experiments, and a low background pressure to minimize competing reactions with contaminants. There
are three sample inlet ports which are trapped with liquid nitrogen and
evacuated to 5×10^{-7} Torr by a Bendix VMW-21 oil diffusion pump. Gaseous , liquid, and solid samples with vapor pressures greater than about
10^{-4} Torr can be used routinely. Each of the three inlet ports has a Varian Associates variable leak valve for controlling the flow rate of
sample addition. The trapped ion analyzer cell is mounted inside a non-
magnetic stainless steel chamber which is pumped through a 6.3 cm diameter throttle valve by a liquid nitrogen trapped 100 l/sec oil diffusion pump. The base pressure in the chamber is normally in the mid-10^{-9}
Torr range following an overnight bakeout at 150°C. During an experiment,
the throttle valve is closed down so that the chamber is pumped at about
20 l/sec, and by adjusting the variable leak valves, steady-state sample
pressures from 10^{-8} to 10^{-3} Torr can be obtained. A Bayard-Albert ioni-

zation gauge is shown in Fig. 5 for measuring pressures in the chamber from 10^{-9} Torr to 10^{-4} Torr. Tubing of 6.3 cm diameter is used to connect the chamber and the ionization gauge in order to minimize the pressure differential. Since the sensitivity of the ionization gauge depends on the composition of the gas in the system, a capacitance manometer (MKS Baratron model 145-AHS-1) is needed. The sensitivity of the ionization gauge for each compound studied is determined by a calibration experiment in the 10^{-5} Torr range. In several hundred experiments of this sort, linear response over one order of magnitude has been observed between the ionization gauge reading and the pressure reading from the capacitance manometer. Measurements of ion-molecule rate constants at various pressures has shown that the linearity and the calibration of the ionization gauge is maintained down to at least 1×10^{-7} Torr. A 1 cm O.D. tube connects the analyzer cell to the capacitance manometer to insure that the actual pressure at the analyzer cell is measured. Both the ionization gauge and the capacitance manometer are coupled directly to the vacuum system so that the calibration procedure can be performed with the analyzer cell in place under actual experimental conditions.

Figure 4 is a schematic diagram of a one-region, trapped ion analyzer cell. Positive or negative ions are produced by electron impact and stored at low energy for several seconds in what is essentially a static magnetic ion trap. Mass analysis of the ions in the trap is accomplished by a detector which excites the cyclotron motion of the ions. For electron impact generation of ions, a beam of electrons is produced from a heated rhenium ribbon filament, collimated by the magnetic field, and accelerated through the analyzer cell by providing a negative bias for the fila-

ment. Emission currents from 10^{-9} to 10^{-6} A are used typically to generate on the order of 10^5 ions inside the analyzer cell. The electron beam
current is pulsed by a grid situated between the filament and the side
plate. The magnitude and width of the electron beam pulse are monitored
as a voltage pulse produced across the 500kΩ impedance to ground of the
electron collector plate. For most experiments, pulsing the electron beam
current is preferable to switching the electron energy above and below
the ionization potential of the compound being studied. A continuous electron beam does allow for feedback regulation but its presence in the analyzer cell distorts the electrostatic fields and can heat and fragment the
trapped ions. A great improvement in the ion trapping efficiency of the
analyzer cell was achieved by placing high line density silver mesh over
the holes where the electron beam passes through the side plates. This
seems to screen out stray electric fields which can perturb the motion
of the ions by penetrating into the analyzer cell from the filament and
the electron collector plate.

The six plates of the analyzer cell are made of OFHC copper which
has been highly polished, plates with silver, and flashed with rhodium.[38]
Laser excitation of the ions stored in the trap and of electron photo-
detachment from negative ions has been accomplished by replacing the solid end plates with high line density screen.[39] The screen seems to work
just as well for trapping ions as the solid end plates. The dimensions
of the analyzer cell and the tolerance on the alignment of the plates
are not of critical importance. Inside dimensions 2.22 x 2.22 x 8.89 cm
are used typically in our laboratory. The analyzer cell is made as wide
as possible, consistent with the magnet gap, and three or four times longer

than wide in order to minimize the the effect of the end plates. The
support rods for the cell plates and the filament block are made of machineable glass ceramic.[40]

The trapped ion analyzer cell serves two important functions in pulsed ICR experiments. First, it is a static ion trap. Ions are trapped
within the analyzer cell by the combined effects of the magnetic field
and an electrostatic potential well established inside the cell by DC
voltages applied to the six metal plates. For positive ions, the side
plates are biased at +1V and the other four plates are set at -0.5V. Negative ions are trapped by reversing the polarity of all the trapping
voltages. The mechanism for ion trapping is similar to the static magnetic ion trap developed by Penning for the study of discharges and used
by Dehmelt and Walls to trap electrons at ultra-low pressures for 3 x 10[6]
sec, or five weeks.[41,42] The motion of ions in a one-region trapped ion
analyzer cell has been discussed in detail by Sharp, Eyler, and Li.[26] In
the direction perpendicular to the magnetic field, ions are constrained to
nearly circular cyclotron orbits with a radius of a few tenths of a milli-
meter. There is also a slow drift motion from one end of the analyzer cell
to the other which is caused by drift of the ions along equipotentials of
the DC electric field inside the cell. The essential purpose of the end
plates is to produce equipotentials which close upon themselves. This prevents ions from drifting out of the analyzer cell or colliding with the
plates. The third fundamental motion of the ions is an oscillatory motion
of frequency

$$\omega_T = \left[\frac{4qV}{md^2}\right]^{1/2} \tag{5}$$

in the direction parallel to the magnetic field.[13] In Eq. 5 , m/q is the mass-to-charge ratio of the ion, V is the applied trapping voltage (the voltage difference between the upper plate and the side plate), and d is the separation between the upper and lower plates. The DC voltages applied to the side plates of the analyzer cell produce an electrostatic potential well which efficiently traps low kinetic energy ions, and the oscillatory motion described by Eq. (5) is due to harmonic-like oscillations in the electrostatic trapping well. One of the nice features of the one-region ICR cell is that the magnitude of the DC voltages applied to the six plates does not significantly alter the performance or the trapping efficiency of the analyzer cell.

Figure 6 shows typical data for the ion trapping efficiency of the

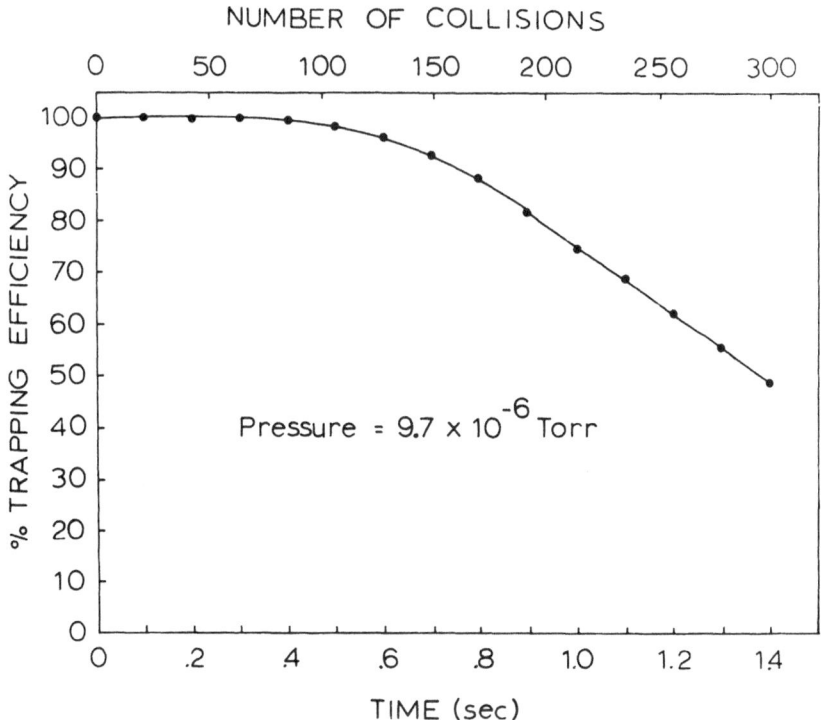

Fig. 6: Ion trapping efficiency of the one-region trapped ICR cell for $C_6H_6^+$ in benzene at 1.2 T and 9.7×10^{-6} Torr.

analyzer cell as a function of reaction time. At an operating pressure of 9.7×10^{-6} Torr and a magnetic field strength of 1.2 T, m/e 78^+ ions are trapped with 50% efficiency for a period of 1.4 sec. This corresponds to about 300 ion-neutral collisions. Diffusive losses perpendicular to the magnetic field are the main factor which limit the trapping efficiency of the analyzer cell.[26] Since the rate of diffusion is inversely proportional to the square of the magnetic field strength, the highest attainable field strength available from the electromagnet is used whenever possible.

The trapped ion analyzer cell also functions as a mass spectrometer. The applied magnetic field, B, causes ions to move in circular cyclotron orbits with a frequency ω given by

$$\omega^2 = \left[\frac{qB}{m}\right]^2 - \frac{4qV}{md^2} , \qquad (6)$$

where qB/m is the classical cyclotron frequency of an ion in a homogeneous magnetic field B, and $4qV/md^2$ is a small correction term due to the effect of the applied trapping voltage V on the motion of the ions.[13] Rearrangement and expansion to first order gives

$$\frac{m}{q} \simeq \frac{B}{\omega} - \frac{2V}{\omega^2 d^2} . \qquad (7)$$

This equation is applied by knowing V and d, measuring the magnetic field strength B with a Hall probe, and detecting the resonance frequency ω with a circuit such as a marginal oscillator. An alternating radio-frequency electric field is produced perpendicular to the magnetic field by connecting the lower plate of the analyzer cell directly to the parallel RLC resonant circuit of the marginal oscillator.[43] When the frequency of the RF electric field is equal to the cyclotron frequency of an ion, the voltage level in the resonant circuit of the marginal oscillator decreases slightly

as the ion is accelerated to larger orbital radius and higher kinetic ener-

gy. This phenomenon is well characterized and provides a sensitive and re-

liable method for detecting positive or negative ions. The resonance fre-

quencies for ions in the analyzer cell can be measured very accurately in

this way, and the mass measurement accuracy of the pulsed ICR spectrometer

is routinely better than ± 0.1 amu.[35,44] Recent experiments have also de-

monstrated that high resolution mass spectra can be obtained with a one-

region trapped ion analyzer cell,[44,45].

A pulsed mode of operation is used for kinetic and equilibrium studies

of ion-molecule reactions. Figure 7 shows the sequence of events for a ty-

pical experiment. The first pulse to be triggered is a pulse of the elec-

tron beam through the interior of the analyzer cell. The bias voltage on

the grid is normally 5V more negative than the filament in order to pre-

vent electrons from entering the analyzer cell. But during the grid pulse

period the bias voltage on the grid is pulsed 5V more positive than the

filament for about 5 msec, thus allowing electrons to enter the analyzer

cell with an energy equal to the voltage difference between the filament

and the side plate. During this short period of time, all primary ions

are formed by electron impact. Either positive or negative ions (depending

on the polarity of the voltages applied to the analyzer cell plates) are

trapped between the side plates by a shallow electrostatic potential well.

Numerous ion-molecule collisions, up to several hundred per ion, occur

during this trapping period.

The second pulse to be triggered is the detect pulse. During the pe-

riod of the detect pulse, ions of a particular mass-to-charge ratio are

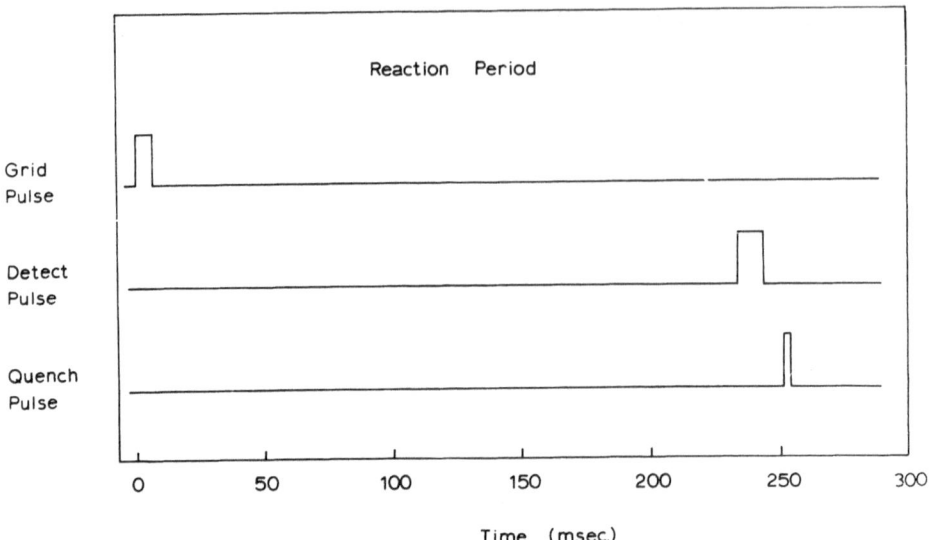

Fig. 7. The sequence of pulses used in pulsed ICR single resonance experiments. The time from the start of the grid pulse to the start of the
detect pulse is called the reaction period. It is also possible to use
a four-pulse sequence where the reactant ions are accelerated prior to
reaction by a double resonance RF pulse.

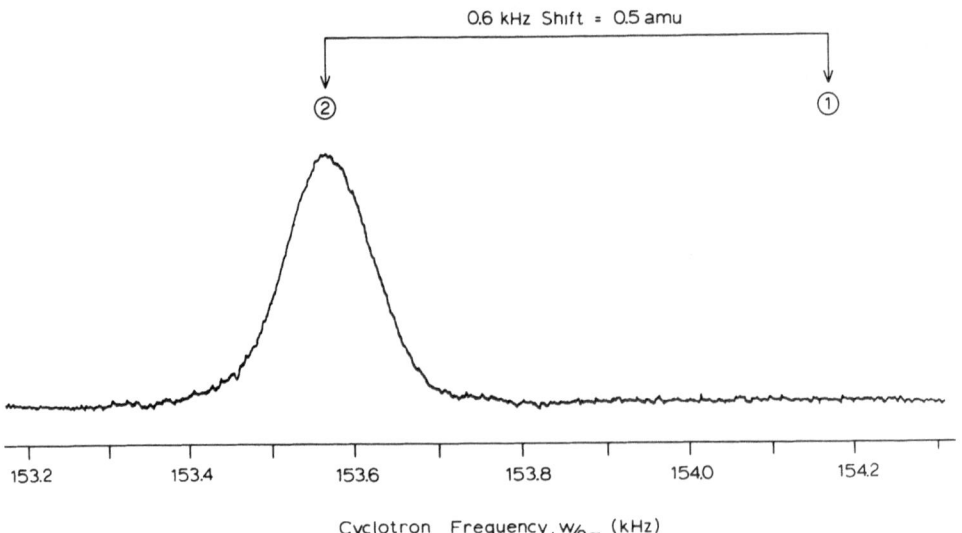

Fig. 8. At a magnetic field strength of 12.8 kG, and a marginal oscillator frequency of $\omega_1 =$ 153.57 kHz, the position of the resonance for
m/e 128 ions can be shifted 0.5 amu by a magnetic field pulse of H' =
50 G or by a change of V' = 1.2 V in the voltage applied to the side
plates of the analyzer cell.

irradiated at their cyclotron frequency by a marginal oscillator, and there
is a pulsed response from the marginal oscillator which is proportional
to the number of resonant ions in the analyzer cell. Proper implementation
of the detect pulse is of critical importance in experiments with a one-
region analyzer cell. During the reaction period (see Fig. 7) the ions must
be allowed to remain at thermal energies. Only during the relatively short
detect pulse period is the kinetic energy of the ions strongly perturbed
by cyclotron resonance irradiation. Three methods have been devised for
experimentally achieving the detect pulse function when using a marginal
oscillator detector: (1) pulsing the strength of the magnetic field,[24] (2)
pulsing the voltage on the side plates,[25,44] and (3) pulsing the oscillation level of the marginal oscillator on and off.[43] The principle of the
first two methods is illustrated in Fig. 8.

With the pulsed magnetic field experiment, the magnetic field strength B
is set off resonance at point 1 during the reaction period so there is
negligible interaction between the marginal oscillator at frequency ω_1
and ions of resonant frequency ω

$$\omega_1 \neq \omega = \left[\left(\frac{qB}{m} \right)^2 - \frac{4qV}{md^2} \right]^{1/2} . \tag{8}$$

There is no response from the marginal oscillator under these conditions,
and the kinetic energy of the ions is not perturbed. However, during the
period of the detect pulse, the magnetic field strength is decreased by
an amount B , causing a shift in the resonant frequency of the ions to
point 2 , which is equal to the frequency of the marginal oscillator,

$$\omega_1 = \omega = \left[\left(\frac{q\,(B - B')}{m} \right)^2 - \frac{4qV}{md^2} \right]^{1/2} . \qquad (9)$$

When the ions are brought into resonance with the marginal oscillator, their kinetic energy is increased by cyclotron resonance power absorption, and a signal is detected at the output of the marginal oscillator.

The second method mentioned above shifts the resonant frequency of the ions by changing the voltage applied to the side plates of the analyzer cell. During the reaction period Eq. (8) applies, and as before, there is negligible interaction between the ions and the marginal oscillator. However, during the period of the detect pulse, the voltage on the side plates is increased by an amount V' in order to shift the resonant frequency of the ions to point 2 ,

$$\omega_1 = \omega = \left[\left| \frac{qB}{m} \right|^2 - \frac{4q\,(V + V')}{md^2} \right]^{1/2} \qquad (10)$$

By modulating the resonant frequency of the ions in this way, their kinetic energy is not perturbed during the reaction period. In practice, pulsing the voltage on the side plates is more convenient than pulsing the magnetic field. Partial differentiation of Eq. (7) at constant frequency

$$\left| \frac{\partial \left| {}^m\!/\!_q \right|}{\partial V} \right|_\omega = \frac{-2}{\omega^2 d^2} \qquad (11)$$

shows that (with conversion of units) a change of 1 V in the voltage applied to the side plates of our analyzer cell causes a shift of 0.42 amu when the marginal oscillator frequency is 153 kHz. Since the linewidths at low pres-

sure are only about 0.1 amu, a 1V pulse applied to the side plates is generally sufficient.

The third method mentioned above for achieving the detect pulse
function involves use of a pulsed marginal oscillator circuit which has
been discussed in detail.[43] The first two methods utilize a marginal oscillator which applies a fixed frequency, fixed amplitude RF voltage to
the lower plate of the analyzer cell while the frequency of the ions is
modulated. The pulsed marginal oscillator, on the other hand, can pulse the
amplitude of the RF voltage applied to the lower plate. There is virtually
no RF irradiation applied to the analyzer cell during the reaction period,
but during the detect period the oscillator turns on to accelerate resonant ions. This has the advantage that the kinetic energy of the ions is
not at all perturbed by the irradiating RF during the reaction period. The
pulsed marginal oscillator is also more useful for experiments at pressures
greater than 10^{-5} Torr because the linewidths are so broadened by ion-neutral collisions that the two frequency modulation methods are not practical.

Following the detect pulse, all the ions, regardless of mass, are rapidly neutralized at the walls of the analyzer cell by the quench pulse.
The trapping action of the cell is destroyed by temporarily inverting the
polarity of the DC voltage applied to the upper plate of the analyzer cell.
A + 10V pulse of 1 msec duration is used to quench positive ions. The quench
pulse stops the reaction and prevents ions from one pulse sequence from over-
lapping into the next sequence. The whole cycle is then automatically repeated. A Newport, Inc. model 700 programmable counter/time is used to measure the frequency of the marginal oscillator, and the duration and delay

time of the pulses. The stability of the timing circuits is on the order
of one part in 10^5 during an experiment.

The theory of cyclotron resonance power absorption can be used to
relate the signal from the marginal oscillator to the number of ions of
a particular mass-to-charge ratio in the analyzer cell. During the detect
pulse period, irradiation at the resonant frequency of an ion produces
a transient signal ΔV which is described by the expression

$$\Delta V = \frac{A}{V_1} \left| \frac{R_f}{1 + GR_f} \right| \tag{12}$$

where A is the power absorbed by the resonant ions, V_1 is the rms voltage
level of the RF applied to the lower plate of the analyzer cell, and G and
R_f are circuit parameters of the marginal oscillator detector.[43] The power
absorption A is the only term which changes in an experiment, and it is
given by

$$A = \frac{Nq^2E^2}{4m} \left| \frac{1 - e^{-\xi t}}{\xi} \right| \tag{13}$$

where N ions of mass-to-charge ratio m/q are irradiated at their resonant
cyclotron frequency, t is time of irradiation, E is the RF electric field
strength. The reduced collision frequency, ξ, accounts for the effect of
non-reaction ion-molecule collisions in damping the coherent cyclotron motion of the ions.[46] In pulsed ICR experiments with a one-region cell the
initial condition t = 0 corresponds to when the detect pulse is triggered,
so Eq. (13) described the full transient response from the marginal oscillator during the period of the detect pulse. The transient signal represented by Eq. (13) is electronically integrated for a time t = τ, usually 7
msec, acquired by a sample-hold circuit, and displayed on the Y-axis of a

recorder. At typical operating pressures on the order of 10^{-6} Torr, it is improbable that the ions suffer ion-molecule collisions during the 7 msec period of the detect pulse. Under these conditions, the limit $\xi t \ll 1$ applies, and the exponential in Eq. (13) can be expanded to give

$$A = \frac{Nq^2E^2t}{4m} .$$
(14)

By keeping the circuit parameters of the marginal oscillator constant, working at constant RF level, and integrating the transient signal for a fixed period, the ICR signals are simply

$$\text{signal} \propto \frac{N}{m} .$$
(15)

Thus, at pressures below 10^{-5} Torr, the marginal oscillator signal is multiplied by the mass m of the ions detected to obtain a value proportional to the number of resonant ions N of a particular mass-to-charge ratio in the analyzer cell.

Ion cyclotron double resonance experiments can be performed with the trapped ion analyzer cell by connecting a second oscillator operating at frequency ω_2 to the upper plate. These experiments, which have been previously described in details, establish the kinetic sequence of reactions and the kinetic energy dependence of the reaction rate constants.[25] Double resonance irradiation is not used when measuring thermal rate constants for ion-molecule reactions.

It is also possible to eject ions or electrons from the analyzer cell by exciting their harmonic oscillatory motion in the direction parallel to the magnetic field.[13] Experimentally this is accomplished by applying a 20mV (rms) RF signal to one of the side plates, the proper frequency being calculated from Eq. (5). The mass resolution of this ion ejection method is far lower than the usual ICR double resonance experiment, but it is very useful in negative ion studies for ejecting low energy electrons which can accumulate in the analyzer cell and produce an excessive space charge.

Development of Frequency Sweep Detectors

Most of the ICR experiments at U.C. Irvine have been done with a one-region analyzer cell, a fixed amplitude marginal oscillator, and the second method discussed above (changing the voltage on the side plates) to shift the resonant frequency of the ions.[44] However, during the last year and a half we have been experimenting with a capacitance bridge detector as a replacement for the marginal oscillator. We have found marginal oscillator circuits to be simple to operate, reliable, and sensitive, but a major limitation is that their frequency cannot be scanned conveniently over a wide range. This means that ICR mass spectra must be recorded by slowly scanning the strength of the magnet, a process which is slow and inflexible. Furthermore, since ion storage efficiency of a trapped ICR cell is proportional to the square of the magnetic field strength, scanning B to obtain a spectrum causes low mass ions to be stored far less efficiently than high mass ions. Another problem we have noticed is that detection of high mass ions is difficult with a marginal oscillator because its intrinsic sensitivity is inversely proportional to mass, Eq. (13), and because practical limitations in circuit design make it difficult to reach the oscillation frequencies useful for ions greater in mass than about m/e 250.

One of our goals lately has been to develop a detector which permits ICR experiments to be done at constant magnetic field strength. In fact the maximum field strength available from the magnet will be used to maximize the ion storage efficiency of the cell and the mass resolution. This permits a mass spectrum to be scanned by sweeping the frequency of the irradiating RF voltage applied to the cell. The ability to rapidly pulse the irradiating RF on and off is also an essential feature for kinetic studies.

Thus far we are encouraged by the progress made in developing a frequency sweep detector. One thing is clear - the electronic circuits required are far more complex than those of a marginal oscillator. Nevertheless, the basis for the method is easily understood and is illustrated in Fig. 9. The right arm of the capacitance bridge consists of an adjustable

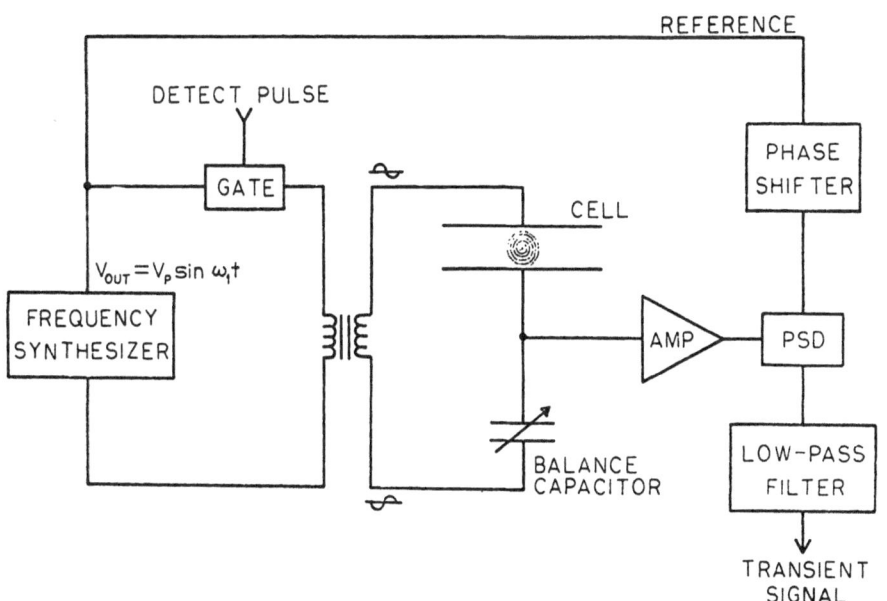

Fig. 9: Block diagram of a capacitance bridge detector for pulsed irradiation and detection of ions stored in a one-region trapped ICR cell.

null capacitor, C_2, and capacitance C_1 formed by the upper and lower plates of an ICR analyzer cell. We are using a one-region cell, but this detection method could be used with all type ICR cell designs as a direct
replacement for the marginal oscillator. Irradiation of the ions is accomplished by passing a RF voltage through a gate (for pulsing) and into a
broadband transformer which has a balanced differential output. In the
absence of resonant ions, C_2 is adjusted equal to C_1 so there is a null
(a virtual ground) at the input of the amplifier. When the bridge is balanced the amplitude V_1 of the irradiating RF can be pulsed on and off
without causing any disturbance at the input of the amplifier. Ions stored in the cell can interact with the irradiating RF when it is pulsed on.
If the resonant frequency of an ion is different from the irradiating frequency, it is not appreciably perturbed and the bridge remains balanced.
However, when resonance is established, an ion is coherently accelerated
to larger radii of gyration as shown by the spiral between the uppper and
lower plates in Fig. 9. The coherent cyclotron motion of the resonant ions
induces "image currents" to flow to the upper and lower plates of the cell.
An RF voltage is developed by the image currents as they pass through the
input impedance of the amplifier . Thus, a "signal" from the capacitance
bridge detector is an RF voltage which appears whenever the irradiating
frequency is close to or the same as the resonant frequency of ions stored
in the cell. Demodulation of the signal with a phase sensitive detector
(PSD) followed by low-pass filtering provides an audio transient such as
shown in Fig. 10. When the irradiating RF is on, there is an increase in
the signal as the ions are accelerated to larger radii of gyration. And
when the RF is turned off, the signal decays exponentially as the ions

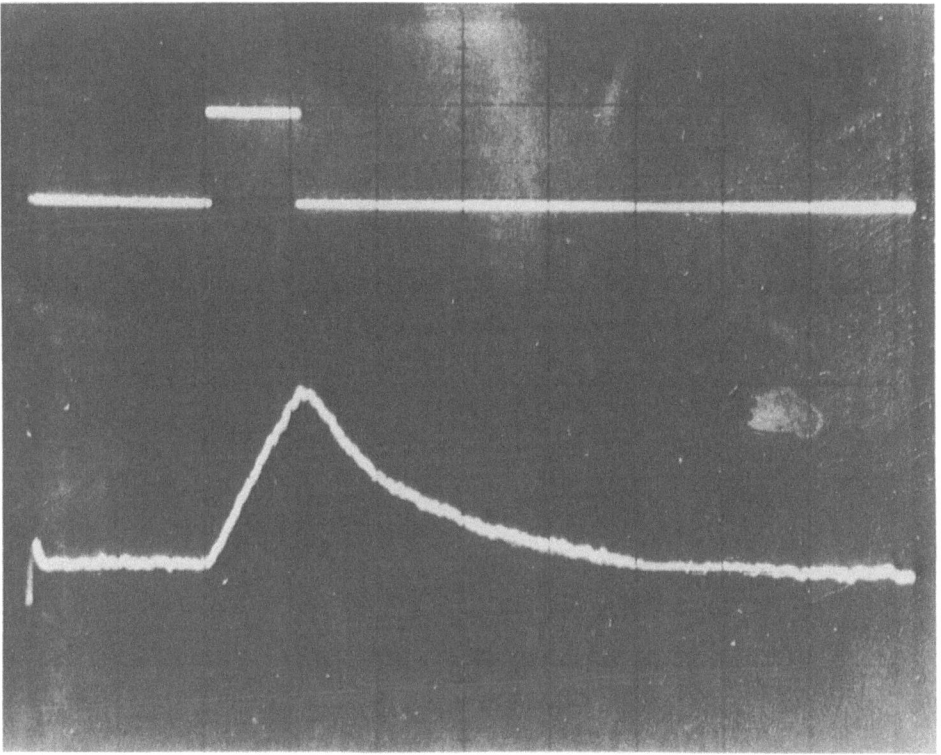

Fig. 10: Oscilloscope traces of an audio transient signal produced by a
capacitance bridge detector (lower trace) and a logic pulse
which turns on the irradiating RF voltage (upper trace). Scale:
5 msec per horizontal division.

emit power to the detector and are slowed down by ion-molecule collisions.

We have been able to operate a capacitance bridge detector from 20 kHz

to 1 MHz to obtain frequency sweep ICR spectra over a wide mass range. This

is obviously the type of detector which will be needed for computer-con-

trolled ICR mass scans using a frequency synthesizer. Furthermore, since

the field strength of superconducting solenoidal magnets cannot be easily

scanned, a detector of this sort is needed. The sensitivity of the capacitance

bridge detector is <u>independent</u> of the mass of the ions detected, and our

prototype circuits have provided far better signal-to-noise than the best of our solid-state marginal oscillators.

Performance of a One-Region Trapped ICR Cell

The ion-molecule reactions in methane are examined in this section to illustrate how experiments with a one-region ICR cell are done. In a typical experiment methane is added through one of the inlet ports to a pressure of 2.5×10^{-6} Torr, and an experiment is initiated by a 3 msec pulse of 30eV electrons through the analyzer cell to form CH_4^{+} (m/e 16) and CH_3^{+} (m/e 15):

$$CH_4 + e^- \begin{cases} \longrightarrow CH_4^{\cdot +} + 2e^- \\ \longrightarrow CH_3^{+} + H\cdot + 2e^- \end{cases} \qquad (16)$$

The positive ions are trapped inside the analyzer cell, the electrons, being of opposite charge, collide within a few microseconds with the side plates, and neutral radicals formed by electron impact or pyrolysis on the hot filament are removed from the cha ber by the vacuum pumps. During the reaction period, the ions react with methane in ways which have been studied previously by a wide variety of techniques:

$$CH_4^{\cdot +} + CH_4 \xrightarrow{k_{13}} CH_5^{+} + CH_3\cdot \qquad (17)$$

$$CH_3^{+} + CH_4 \xrightarrow{k_{14}} C_2H_5^{+} + H_2 \qquad (18)$$

The date in Fig. 11 show mass spectra of this system taken at various delay times. In Fig. 11(a) the detect pulse delay time is net at 5.2 msec (2.2 msec after the 3 msec grid pulse is turned off), the frequency of the pulsed marginal oscillator is set at 307 kHz, and the magnetic field strength was

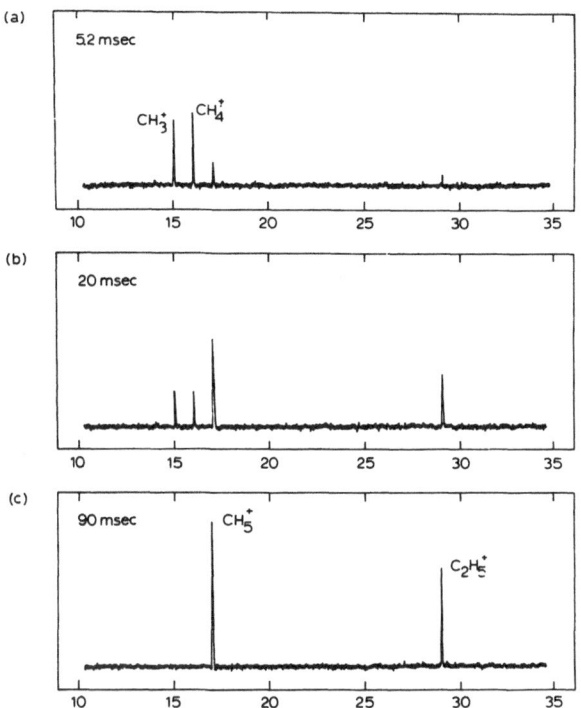

Fig. 11: Pulsed ICR mass spectra recorded at various detect pulse delay times: (a) 5.2 msec, (b) 20 msec, and (c) 90 msec. Ions in methane at 2.5 x 10⁻⁶ Torr are observed at a marginal oscillator frequency of 307 kHz.

scanned one time from 2kG to 7kG in a period of 10 minutes. After such a short delay time, the main ions present in the analyzer cell are the electron impact products CH_3^+ and CH_4^+. Fig. 11 (b) was recorded in the same way except the detect pulse delay time was set at 20.0 msec. The intensity of the reactant ions is seen to be lower due to a significant increase in the product ions CH_5^+ and $C_2H_5^+$. Finally, Fig. 11 (c) shows that after 90 msec reactions 17 and 18 have gone to completion, and the product ions CH_5^+ and $C_2H_5^+$ are the only ones in the mass spectrum. The peak heights in mass spectra of this type are only semi-quantitatively significant because the mass resolution is so high that the X-Y recorder does not have time to trace accurately the full peak height of the signal. Slower scan rates over each peak do produce signals described quantitatively by Eq. (15).

A second mode of operation from the pulsed ICR mass spectrometer generates data, such as Fig. 12, in the form of "time plots." The frequency of the marginal oscillator is fixed at 307 kHz, and the magnetic field strength is adjusted to be in resonance during the detect pulse period with an ion of chosen m/e. A scan generator varies the delay time of the detect pulse to produce a plot of X-Y recorder of ion abundance vs. reaction time. Figure 12 (a) shows that the abundance of CH_4^+ increases rapidly during the 3msec period of the electron beam pulse, but then decreases due to reaction with methane. Figure 12(b) shows a zero initial abundance of CH_5^+ and an exponential increase over a period of about 80 msec due to reaction 17. These observations can be made more quantitative by considering the kinetic rate equations for this reaction system. Since the concentration of CH_4^+ in the analyer cell (about 10^5 particles $\cdot$ cm^{-3}) is much less than the concentration of methane (8.0×10^{10} particles $\cdot$ cm^{-3}), pseudo - first order kinetics are followed. Thus, during the reaction period, the differential rate equation for CH_4^+ is

$$\frac{d\ [CH_4^+]}{dt} = -k_{17}\ [CH_4]\ [CH_4^+]\ . \tag{19}$$

This can be integrated to give

$$[CH_4^+] = [CH_4^+]_o \exp\ (-k_{17}\ [CH_4]), \tag{20}$$

where $[CH_4^+]_o$ is the initial concentration of CH_4^+ in the analyzer cell

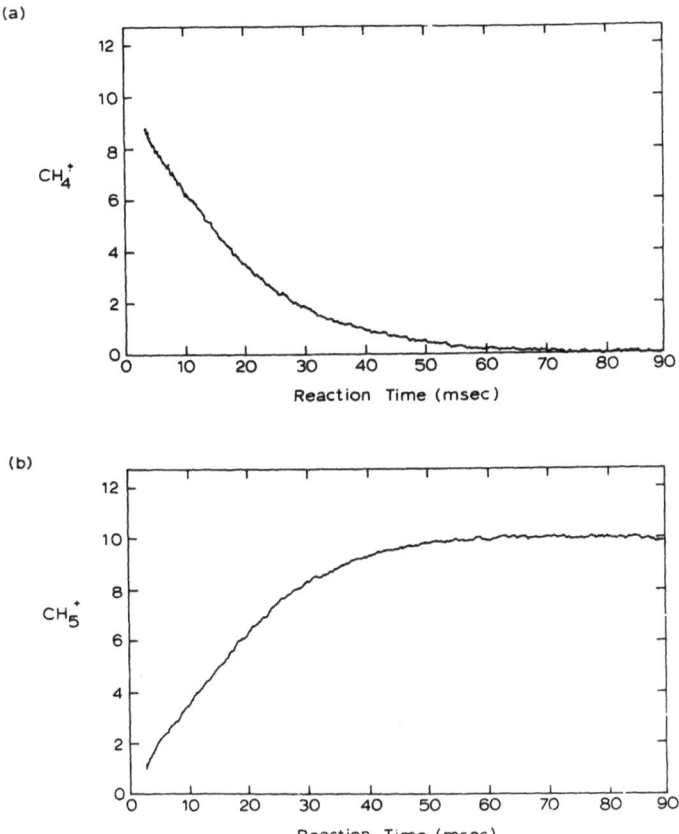

(a)

CH₄⁺

Reaction Time (msec)

(b)

CH₅⁺

Reaction Time (msec)

Fig. 12: Pulsed ICR time plots of $CH_4^{+\cdot}$ (m/e 16) and CH_5^+ (m/e 17) at a methane pressure of 2.5 x 10⁻⁶ Torr.

after the electron beam has been turned off, k_{17} is the bimolecular rate constant for reaction 17, and t is reaction time. In this experiment the reaction time is equal to the detect pulse delay time. Similarly, the differential rate equation for CH_5^+,

$$\frac{d\ [CH_5^+]}{dt} = k_{17}\ [CH_4]\ [CH_4^{\ \dot{+}}] \tag{21}$$

can be integrated by substitution of Eq. (20) for $[CH_4^{\ \dot{+}}]$ to give

$$[CH_5^+] = [CH_5^+]_\infty \left(1 - \exp\ (\ -\ k_{17}\ [CH_4]\ t)\right), \tag{22}$$

where $[CH_5^+]_\infty$ is the final concentration of CH_5^+. Semilogarithmic plots of these data give $k_{17} = 1.1 \times 10^{-9}\ cm^3 \cdot molecule^{-1} \cdot sec^{-1}$. This value is in good agreement with previously published values.[47-49]

The high ion trapping efficiency of the one-region ICR cell allows experiments to be run at far lower pressures than is possible with most other instruments used for studying ion-molecule reactions. Reactions observed in a chemical ionization mass spectrometer in a few microseconds at pressures of 0.1 Torr are observed in tenths of a second in a pulsed ICR spectrometer operated at a pressure of 10^{-6} Torr. In studying bimolecular reactions there doesn't seem to be much difference between the various methods, but termolecular and higher order reactions are difficult to observe in a pulsed ICR spectrometer.

Reactions 17 and 18 are both so exothermic that the rate of the reverse reaction is negligible, but numerous examples have been found to reversible ion-molecule reactions.[28-31] One such reaction is the reversible proton transfer reaction involving the negative ions CH_3O^- and $C_2H_5O^-$,

$$CH_3O^- + C_2H_5OH = C_2H_5O^- + CH_3OH \quad . \quad (23)$$

At pressures on the order of 10^{-6} Torr, the relative abundance of $CH_3O^-/$
$C_2H_5O^-$ reaches a constant value after a reaction period of a few tenths of
a second, indicating the attainment of equilibrium. The equilibrium constant for reaction 23 is evaluated by using an ionization gauge calibrated
by a capacitance manometer to measure the partial pressures of the two alcohols, and by using the pulsed ICR time plots to measure the steady-state
relative abundance of the ions CH_3O^- and $C_2H_5O^-$. An extensive discussion
of the equilibrium pulsed ICR experiments is in preparation.[50]

One of the features of the pulsed ICR experiments is that the reaction period is far longer than the times needed to generate and mass ana-
lyze the ions. The grid pulse for firing the electron beam through the analyzer cell is typically 5 msec long, and the detect pulse for mass analy-
zing the ions in the analyzer cell is typically 7 msec long. In comparison, the reaction period is typically from 100 msec to 1 sec long, depending on the pressure of the reactants and the rate of the ion-molecule
reaction being studied. This situation greatly simplifies the analysis
of the data because the pressure can be adjusted to insure that the reaction proceeds to a negligible extent during the grid pulse and detect pulse
periods.

Conclusion

The full range of ICR experiments can be performed with a one-region trapped ion analyzer cell by dispersing various events in "time" rather than in "space". Earlier drift cell designs had one region for formation of ions by electron impact, a second region for mass analysis, and a third region for collection of the total ion current. It really isn't necessary to have a separate region of the analyzer cell for each experimental function because all functions can be performed in a one-region analyzer cell by various pulsing sequences. This not only makes for simpler operation, but also provides a number of significant performance advantages. For example, the mass resolution of drift-type ICR analyzer cells has been shown to be limited by the residence time of the ions in the analyzer region.[51] The high ion trapping efficiency of the one-region analyzer cell eliminates this limitation and allows for greatly improved mass resolution.[44] When Fourier Transform and capacitance bridge detectors are used with a one-region analyzer cell, truly high resolution ($M/\Delta M > 10^4$) performance can be achieved.[36,37] The versatile trapped ion cell developed by McMahon and Beauchamp is capable of high ion trapping efficiency,[27] but its mass resolution is limited by the residence time in the analyzer region in the same way as the drift-type analyzer cell.

References

1. J. A. Hipple, H. Sommer, and H. A. Thomas, Phys. Rev. 76, 1877 (1949)

2. H. Sommer, H. A. Thomas, and J. A. Hipple, Phys. Rev. 82, 697 (1951)

3. D. Alpert and R. S. Buritz, J. Appl. Phys. 25, 202 (1954)

4. D. Lichtmann, J. Appl. Phys. 31, 1213 (1960)

5. R. W. Lawson, J. Sci. Instrum. 39, 281 (1962)

6. H. B. Niemann and B. C. Kennedy, Rev. Sci. Instrum. 37, 722 (1966)

7. J. L. Beauchamp, L. R. Anders, and J. D. Baldeschwieler, J. Am. Chem. Soc. 89, 4569 (1967)

8. J. D. Baldeschwieler and S. S. Woodgate, Accounts Chem. Res. 4, 114 (1971)

9. D. Wobschall, Rev. Sci. Instrum. 36, 466 (1965)

10. D. Wobschall, J. R. Graham, Jr., and D. Malone, Phys. Rev. 131, 1565 (1963)

11. D. Wobschall, J. R. Graham, Jr., and D. Malone, J. Chem. Phys. 42, 3955 (1965)

12. D. Wobschall, R. A. Fluegge, and J. R. Graham, Jr., J. Chem. Phys. 47, 4091 (1967)

13. J. L. Beauchamp and J. T. Armstrong, Rev. Sci. Instrum. 40, 123 (1969).

14. T. E. Sharp, J. R. Eyler, and E. Li, Int. J. Mass Spectrom. Ion Phys., 9, 421 (1972)

15. T. F. Knott and M. Riggin, Can. J. Phys. 52, 427 (1974)

16. A recent book on this subject is: T. A. Lehman and M. M. Bursey, "Ion Cyclotron Resonance Spectrometry," Wiley, New York, 1976

17. R. P. Clow and J. H. Futrell, J. Am. Chem. Soc. 94, 3748 (1972)

18. R. P. Clow and J. H. Futrell, Int. J. Mass Spectrom. Ion Phys. 4, 165 (1970)

19. R. P. Clow and J. H. Futrell, Int. J. Mass Spectrom. Ion Phys. 8, 119 (1972)

20. T. B. McMahon and J. L. Beauchamp, Rev. Sci. Instrum. 42, 1632 (1971)

21. M. T. Bowers, D. D. Elleman, and J. L. Beauchamp, J. Phys. Chem. 72, 3599 (1968)

22. A. G. Marshall and S. E. Buttrill, Jr., J. Chem. Phys. 52, 2752 (1970)

23. M. B. Comisarow, J. Chem. Phys. 55, 205 (1971)

24. R. T. McIver, Jr., Rev. Sci. Instrum. 41, 555 (1970)

25. R. T. McIver, Jr. and R. C. Dunbar, Int. J. Mass Spectrom. Ion Phys. 7, 471 (1971)

26. T. E. Sharp, J. R. Eyler, and E. Li, Int. J. Mass Spectrom. Ion Phys. 9, 421 (1972)

27. T. B. McMahon and J. L. Beauchamp, Rev. Sci. Instrum. 43, 509 (1972)

28. M. T. Bowers, D. H. Aue, H. M. Webb, and R. T. McIver, Jr., J. Am. Chem. Soc. 93, 4314 (1971)

29. R. T. McIver, Jr. and J. R. Eyler, J. Am. Chem. Soc. 93, 6334 (1971)

30. R. T. McIver, Jr. and J. S. Miller, J. Am. Chem. Soc. 96, 4323 (1974)

31. R. W. Taft in "Proton Transfer Reactions" (E. F. Caldin and V. Gold, Editors) Chapman and Hall, Ltd., London (1975)

32. J. I. Brauman, W. N. Olmstead, and C. A. Lieder, J. Am. Chem. Soc. 96, 4030 (1974)

33. W. N. Olmstead and J. I. Brauman, J. Am. Chem. Soc. 99, 4219 (1977)

34. R. T. McIver, Jr., E.B. Ledford, Jr., and J. S. Miller, Anal. Chem. 47, 692 (1975)

35. E. B. Ledford, Jr. and R. T. McIver, Jr., Int. J. Mass Spectrom. Ion Phys. 22, 399 (1976)

36. R. L. Hunter and R. T. McIver, Jr., Chem. Phys., Lett., in press (1977)

37. M. B. Comisarow and A. G. Marshall, J. Chem. Phys. 64, 110 (1976)

38. The prescription used for plating the copper is as follows: high gloss silver plate, .001" per QQ-S-365 followed by .0001" rhodium flash per MIL-R-46085.

39. J. R. Eyler, Rev. Sci. Instrum. 45, 1154 (1974)

40. Corning Glass type 9658 machineable glass ceramic was used for the insulating support rods of the analyzer cell.

41. H. G. Dehmelt and F. L. Walls, Phys. Rev. Lett. 21, 127 (1968)

42. F. L. Walls and G. H. Dunn, Phys. Today 27, 30 (1974)

43. R. T. McIver, Jr., Rev. Sci. Instrum. 44, 1071 (1973)

44. R. T. McIver, Jr. and A. D. Baranyi, Int. J. Mass Spectrom. Ion Phys. 7, 471 (1971)

45. M. B. Comisarow and A. G. Marshall, J. Chem. Phys. 62, 293 (1975)

46. M. B. Comisarow, J. Chem. Phys. 55, 205 (1971)

47. F. H. Field, J. L. Franklin, and M. S. B. Munson, J. Am. Chem. Soc. 85, 3575 (1963)

48. S. K. Gupta, E. G. Jones, A. G. Harrison, and J. J. Myher, Can. J. Chem. 45, 3107 (1967)

49. J. H. Futrell, T. O. Tiernan, F. P. Abramson, and C. D. Miller, Rev. Sci. Instrum. 39, 340 (1968)

50. J. E. Bartmess, J. S. Miller, and R. T. McIver, Jr., in prep.

51. S. E. Buttrill, Jr., J. Chem. Phys. 50, 4125 (1969)

Fourier Transform Ion Cyclotron Resonance Spectroscopy

M. Comisarov
Chemistry Department
University of British Columbia
Vancouver, B. C. , Canada V6T 1W5

In recent years several spectroscopic techniques have been revolutionized by the introduction of Fourier multiplex methods. The principal advantage of the Fourier method is that the whole spectrum may be obtained in a very short period of time, namely the amount of time which a conventional spectrometer would require to observe just a single point in the spectrum. The most important applications of Fourier methods have been to nuclear magnetic resonance spectroscopy and to infra-red spectroscopy, but the methods have also been found useful in optical spectroscopy, microwave spectroscopy and even electrochemistry. The advantages of the Fourier method would clearly be desirable in the field of mass spectrometry and in principle can be realized by the development of the Fourier transform ion cyclotron resonance (FT-ICR)[1-5] mass spectrometer.

Conventional ion cyclotron resonance (ICR) mass spectrometers yield mass spectra by sequentially exciting the cyclotron motion of the sample ions and simultaneously detecting that excited cyclotron motion, one mass at a time. Typically, about 20 minutes are required to obtain a mass spectrum of reasonable signal to noise ratio over the mass range 15-200. In principle, the FT-ICR spectrometer can yield the whole mass spectrum in the time (typically a few seconds) which a conventional ICR spectrometer would require to scan through just a single line in the mass spectrum by operating in the following manner: The spectrometer excites the cyclotron motion of all ions in the sample at once and then stores the time domain transient signal which results from that excited cyclotron motion. Since

the time domain signal contains contributions from all ions whose motion

has been excited, the frequency domain signal, which is obtained by nume-

rical Fourier transformation of the time domain signal, will contain the

whole ion cyclotron resonance mass spectrum. By accumulation of a number

of time domain signals prior to Fourier transformation, the sensitivity

of measurement can be improved in proportion to the square root of the

number of accumulated transients.

Figure 1 is a block diagram of a prototype Fourier transform ion

cyclotron resonance (FT-ICR) spectrometer. The ICR cell is a trapped-ion

cell of conventional design[6] which is placed in the magnetic field of a

high-field electromagnet. Ions are formed in the cell by short electron

beam pulse from an electron gun (not shown) and are constrained by the

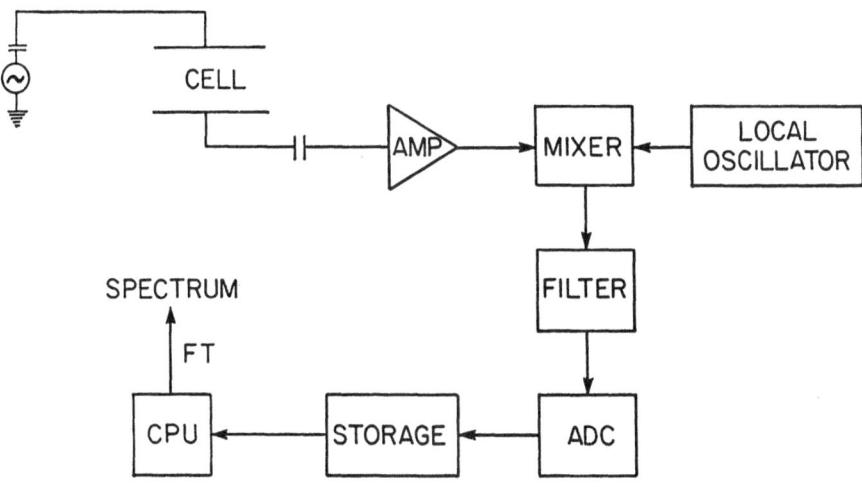

Fig. 1. Prototype Fourier transform ion cyclotron resonance (FT-ICR) spectrometer. The operation of this spectrometer is described in the text.

magnetic field to circular orbits; the angular frequency, ω, of the ion
motion given by the well-known cyclotron equation

$$\omega = qB/m \tag{1}$$

Excited ion cyclotron motion is then achieved by turning on the oscillator which is connected to the top plate of the ICR cell and rapidly sweeping the oscillator over the desired frequency range. The oscillator is
then turned off. The ions whose ICR motion has been excited will then proceed on their orbital paths with angular frequencies which are characteristic of their masses, according to equation (1). In the absence of ion-
neutral collisions (zero pressure limit) and ion loss processes this excited ICR motion will continue indefinitely. A signal is induced by the excited ICR motion in the bottom plate of the ICR cell, then amplified by a
broad-band (10 kHz - 3MHz) amplifier (AMP). This signal is a sum of the
alternating signals resulting from excited ICR motion at each ion mass. The
amplified signal from the amplifier and a signal from a local oscillator
are sent to a mixer, the output of which is passed through a low pass fil-
ter before digitization by an analog-to-digital converter (ADC). The digi-
tized signal is then stored in a digital computer. A quench pulse[6] is then
applied to the trapped ion cell and the cycle repeated for a chosen number
of times. Fourier transformation of the summed, stored time-domain signal
will give the ICR frequency spectrum.

Figure 2 shows an example of an FT-ICR time domain signal which was
obtained by the excitation, mixing, filtering, digitizing and storing
procedure described above. This time domain signal was obtained[4] by excitation of a sample of bromoadamantane ions formed by electron impact on
dibromoadamantane. Fourier transformation of this time domain signal re-

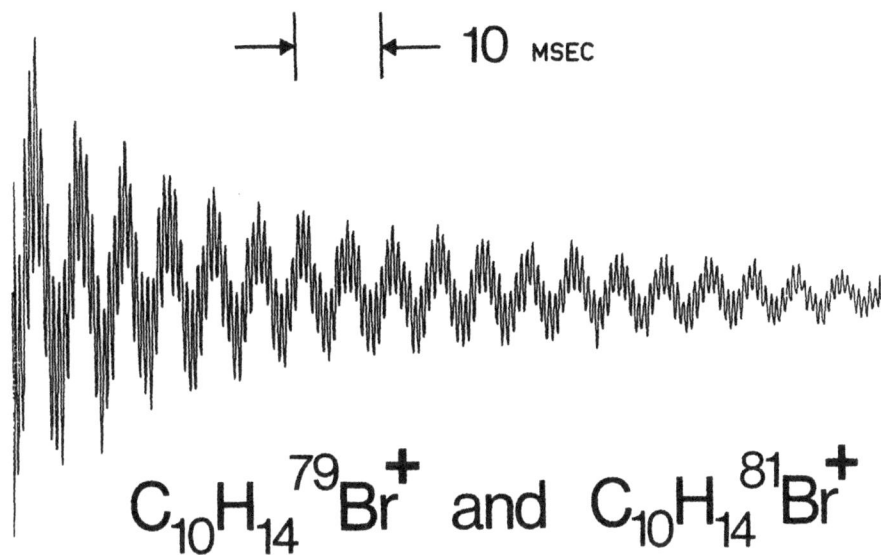

Fig. 2. A single time domain transient cyclotron resonance signal from an excited sample of bromoadamantane[4].

sults in the FT-ICR frequency domain spectrum shown in Figure 3. Another example of a FT-ICR spectrum is shown in Figure 4. This spectrum is noteworthy because it shows that the mass range of the FT-ICR technique extends to over 1000 amu. In contrast, the mass range of conventional ICR spectrometers is usually limited to about 200 amu.

Ion cyclotron resonance spectrometers have in the past been considered to be low to medium resolution mass spectrometers. The FT-ICR spectrometer however has potential as a high or even ultrahigh resolution instrument. Figure 5 shows an FT-ICR ultrahigh resolution mass spectrum near m/e=28 of the molecular ions in a ternary mixture of nitrogen, ethylene and carbon

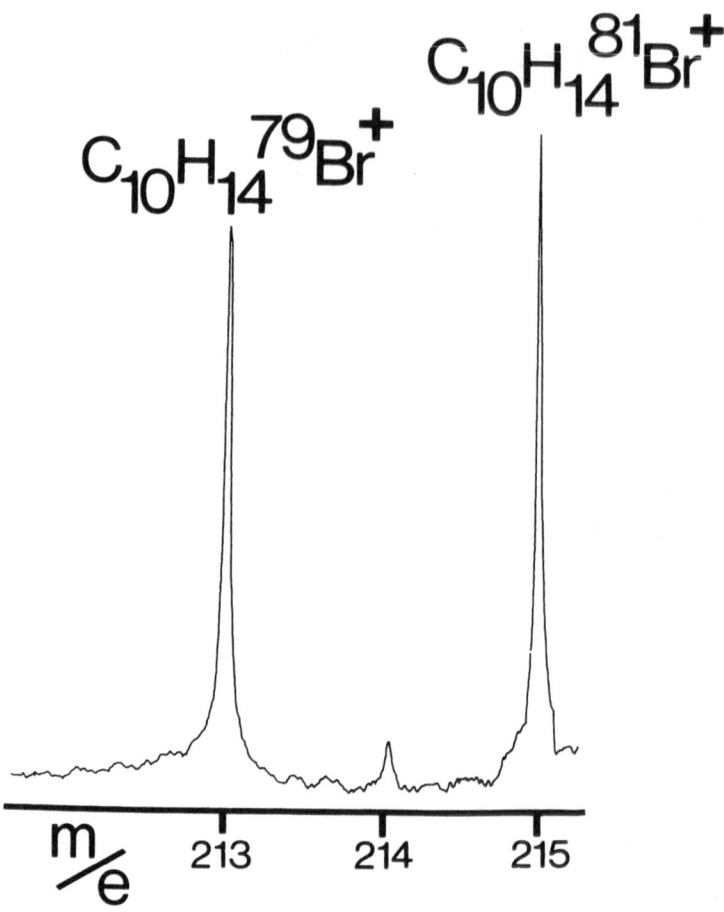

Fig. 3. FT-ICR mass spectrum obtained by numerical Fourier transformation of the time domain signal of Figure 2.

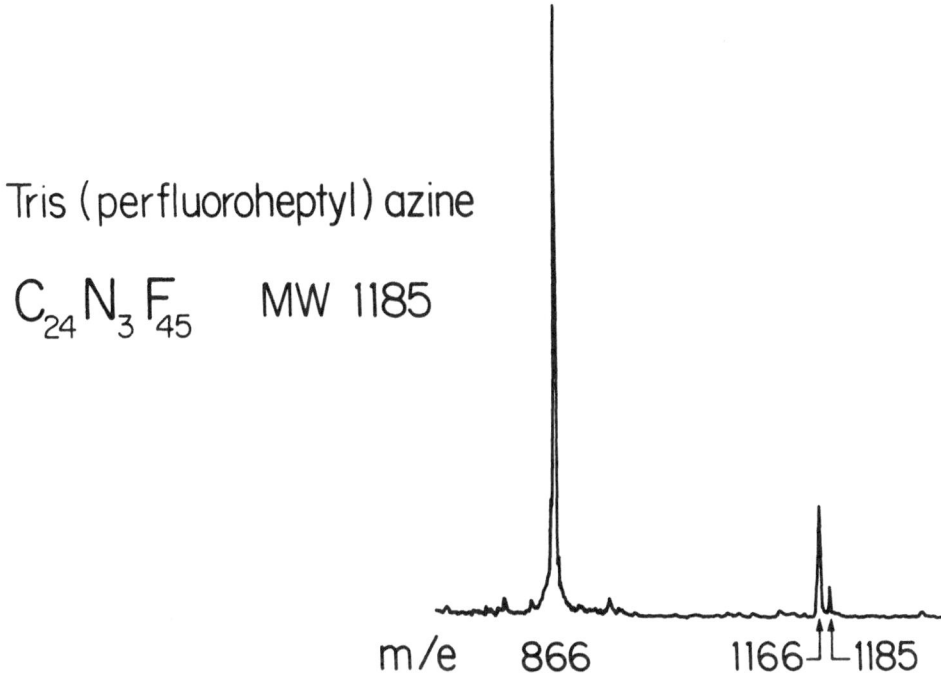

Fig. 4. FT-ICR mass spectrum from m/e 850 to m/e 1200 of Tris(perfluoro-heptyl) azine.

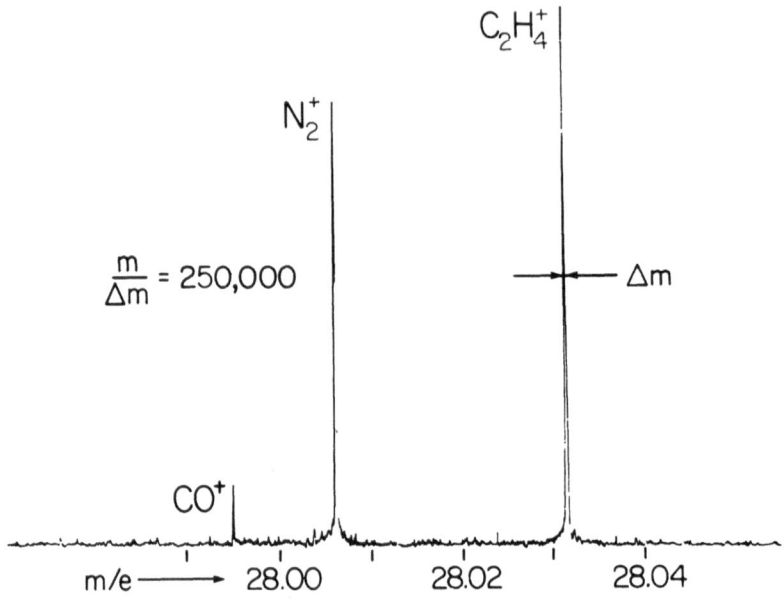

Fig. 5. Ultrahigh resolution FT-ICR mass spectrum of a ternary mixture of nitrogen, ethylene and carbon monoxide near m/e 28.

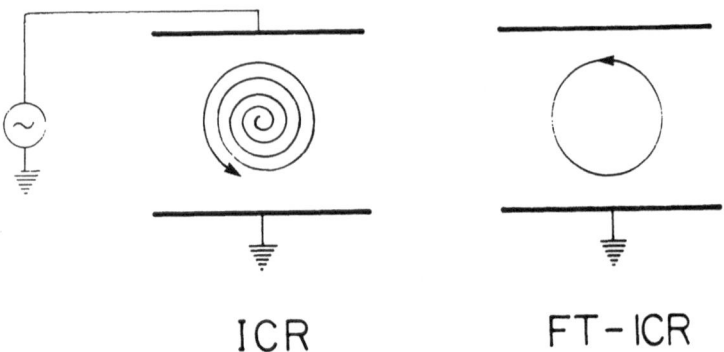

Fig. 6. Ion motion during the detection period for a conventional ICR spectrometer and for a FT-ICR spectrometer. For the conventional ICR spectrometer the ion motion is excited by the applied radiofrequency electric field and the presence of the excited ion motion is detected by noting the power absorption from a radiofrequency oscillator of the marginal oscillator type. For the FT-ICR experiment, the ion motion is excited in a spiral path as in the conventional ICR experiment but the excited ion motion is detected after the radiofrequency oscillator is turned off. In the absence of ion molecule collisions and other processes which can interrupt the coherent motion of the excited ions, the FT-ICR motion will be as shown above. The coherent ion motion induces a signal in the plates of the ICR cell which is representative of the excited ion motion.

monoxide. One feature of this experiment which deserves comment is that the mass resolution of one part in 250,000 shown in Figure 5 was obtained with a magnet whose homogeneity was only one part in 9,000 over volume of the trapped ion ICR cell. The reason why it is possible to obtain a mass resolution in the FT-ICR experiment which is superior to that obtained in the conventional ICR experiment, when the same magnet and the same trapped-ion ICR cell is used for both experiments, appears to be related to the temporal separation between the cyclotron resonance excitation and detection processes which exists in the FT-ICR experiment. In the conventional ICR experiment, ions are excited by a radiofrequency electric field applied across the ICR cell and execute the spiral motion shown in Figure 6. The detection of the excited cyclotron motion is achieved by measuring the power absorption of the ions from the radio frequency electric field. This detection is readily achieved with a marginal oscillator type of radiofrequency oscillator. Thus in the conventional ICR experiment ion excitation and ion detection are simultaneous. In the FT-ICR experiment ion detection is achieved by measuring the time domain signal induced in the plates of the ICR cell by the excited ion motion after the ion motion has been coherently excited. In the absence of ion-molecule collisions and other processes which interupt the coherent ion motion, the excited ICR motion will be as shown in Figure 6. It is clear from Figure 6 that during the detection period the excited ions in the conventional ICR experiment will experience the whole magnetic field over the volume of the spiral path shown in Figure 6 . In contrast, during the detection period the excited ions in the FT-ICR experiment will experience a more restricted region of the magnetic field and therefore a more limited range of magnetic inhomogeneities. For example, in the hypothetical case that the magnetic field inhomogeneity consisted solely of a radial gradient from the center of the ICR cell, (i.e. the magnetic field increased or decreased from the center of the ICR cell), the ion in a conventional ICR experiment would experience the entire magnetic inhomogeneity during its detection period, whereas the ion in the FT-ICR experiment would experience a constant magnetic field during its detection period. Another factor which should be considered is that the FT-ICR motion shown in Figure 6 may effectively average out the magnetic field inhomogeneities experienced by the ion via a process which is analogous to the technique of "spinning the sample"[7] which is used in high resolution nuclear magnetic resonance (NMR) spectroscopy to allow the NMR sample to see a magnetic field which is more homogeneous than if the sample were not spun. This effect will apply to the FT-ICR motion shown in Figure 6 because the ion retraces the same path on each cycle but will not apply to the conventional ICR motion shown in Figure 6 because the average magnetic field seen by the ion changes as the radius of the ion motion is increased during the detection period.

While the above results must be regarded as prototype experiment which only serve to indicate some of the features of the FT-ICR method, the results are encouraging and of sufficient interest to warrant further exploitation of the FT-ICR technique.

Acknowledgement - This research was supported by the National Research Council of Canada.

References

1.   M. Comisarow and A. G. Marshall, Chem. Phys. Lett. 25, 282 (1974).

2.   M. Comisarow and A. G. Marshall, Chem. Phys. Lett. 26, 489 (1974).

3.   M. Comisarow and A. G. Marshall, Can. J. Chem. 52, 1997 (1974).

4.   M. Comisarow and A. G. Marshall, J. Chem. Phys. 62, 293 (1975).

5.   M. Comisarow and A. G. Marshall, J. Chem. Phys. 64, 110 (1976).

6.   R. T. McIver, Jr. Rev. Sci. Inst. 41, 555 (1970).

7.   F. Bloch, Phys. Rev. 94, 596 (1954).

Mechanistic Studies of Some Gas Phase Reactions of $O^{\cdot-}$ Ions with Organic Substrates

J. H. J. Dawson and N. M. M. Nibbering
Laboratory of Organic Chemistry, University of Amsterdam
Nieuve Achtergracht 129, Amsterdam, The Netherlands

Introduction

Recently we published a paper describing the results of our research under the $O^{\cdot-}/CH_3C\equiv CH$ system. [1] By working with the labelled
propyne $CD_3C\equiv CH$ it was shown that the principle cross ions, $C_3H_2^{-\cdot}$
and $C_3H_3^-$ each exist with two non-isomerising structures, i.e. $H\dot{C}=C=\bar{C}H/$
$\bar{C}H_2-C\equiv C^-$ and $\bar{C}H_2-C\equiv CH/CH_3-C\equiv C^-$. (It has recently been shown that $C_3H_3^+$
ion exists in two stable structures:

$$HC\overset{\underset{\textstyle CH}{\diagup\diagdown}}{\underset{\textstyle}{\quad}}CH \quad \text{and} \quad HC\equiv\overset{+}{C}-CH_2 \;.^{2})$$

During the study questions raised concerning the precise nature of the
mechanism of the interaction between $O^{-\cdot}$ and the organic molecule. Of
particular interest is the question of how the $\dot{C}H_2-C\equiv C^-$ ion can be formed, since it requires that $O^{-\cdot}$ shall abstract hydrogen nuclei from
both ends of the propyne molecule.

For this and for the other reactions three mechanisms can be discussed:

(i)   Close range H^+ and/or $H^\cdot$ transfer to $O^{-\cdot}$.

(ii) The formation of a loosely bonded complex, akin to the classic
      "orbiting complex" held together by electrostatic forces or by
      hydrogen bonds, in which a reaction could proceed in several steps
      as the colliding particles move around each other.

(iii) The formation of a chemically bonded intermediate complex such as
      would be postulated in organic chemistry. The complex would under-

go logical internal rearrangements and would then decompose to give the product ions.

There can be little doubt that so far as ions such as $CH_3C\equiv C^-$ are concerned, the first of these mechanisms is operative though not necessarily to the exclusion of the others. This is evidenced by the almost invariable observation of a large positive double resonance signal (see Figure 1) for the reaction

$$O^{-\bullet} + M \longrightarrow OH^{\bullet} + (M - H)^- \qquad (1)$$

where M is an organic molecule containing some acidic hydrogen(s). Under the conditions of excitation by double resonance irradiation the lifetimes of the complexes required for the second and third mechanisms are reduced

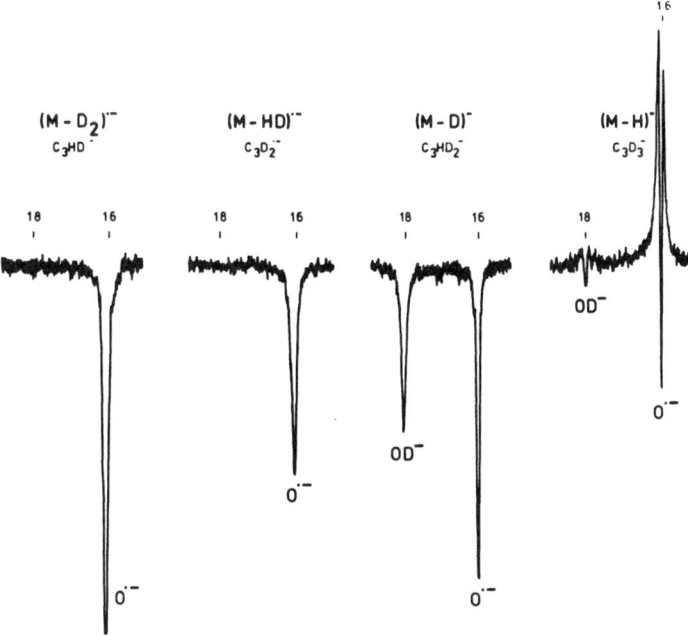

Fig 1. Double resonance spectra recorded on the ions m/e 39, 40, 41 and 42 in a 1:1 mixture of N_2O and $CH_3C\equiv CH$ at a total pressure of $\approx 3\times10^{-5}$ Torr.

so that reactions which proceed through the first mechanism become more

favourable. For the formation of $\overset{\bullet}{C}H_2-C\equiv C^-$ the second and third reaction

mechanisms may be crystallised in the form of equations 2 and 3 shown

below:

$$O^{-\bullet} + CH_3C\equiv CH \longrightarrow [OH^\bullet + CH_3C\equiv \bar{C}]^* \longrightarrow H_2O + CH_2-C\equiv C^- \qquad (2)$$

orbiting complex

$$O^{-\bullet} + CH_3C\equiv CH \longrightarrow [CH_3-C=\overset{\bullet}{C}H \rightarrow CH_3-C=\overset{\bullet}{C}^-]^* \longrightarrow H_2O + \overset{\bullet}{C}H_2-C\equiv C^- \qquad (3)$$
$$\phantom{O^{-\bullet} + CH_3C\equiv CH \longrightarrow [}O^- OH$$

chemically bonded complex

There is evidence from the work of Harrison and Jennings[3] to the effect

that the ion $CH_3-C(O^-)C=CH^\bullet$ can exist as a long lived radical anion.

Unfortunately it does not seem possible to investigate this system

any further by ICR alone, and so we have extended our work to include

an investigation of reactions of a number of oxygen containing molecules.

Then, by the use of ^{18}O labelled reactant anions it becomes possible to

probe more deeply into the mechanistic details of the reactions which

occur.

Results and discussion

Acetone and dimethylsulphoxide

It is clear that the ractions of $O^{-\bullet}$ [4] with acetone provide a key

system for study with labelled $^{18}O^{-\bullet}$. The cross products of the reac-

tions in the unlabelled system have been found by Harrison and Jennings[3]

to be OH^-, $CH_3COCH^{-\bullet+}$, $CH_3COCH_2^-$ and CH_3COO^- ($\sim$ 1% of total yield)

[1:1 mixture with nitrous oxide at a pressure in the 10^{-5} Torr range where

$CH_3COCH_2^-$ is also a product of OH^-].

[†] We use here the structure assigned to the m/e 56 $(M-H_2)^{-\bullet}$ ion by

Harrison and Jennings.[3]

The HC≡CO⁻ ion is taken to be formed exclusively by the loss of a methyl radical from $CH_3COCH^{-\bullet}$. It is surprising that the reaction

$$0^{-\bullet} + CH_3COCH_3 \longrightarrow CH_3COO^- + CH_3^\bullet \tag{4}$$

accounts for only about 1% of the overall yield since a comparison with
conventional solution chemistry[5] would lead one to expect that the formation of the tetrahedral intermediate[6] complex I would be a more significant process.

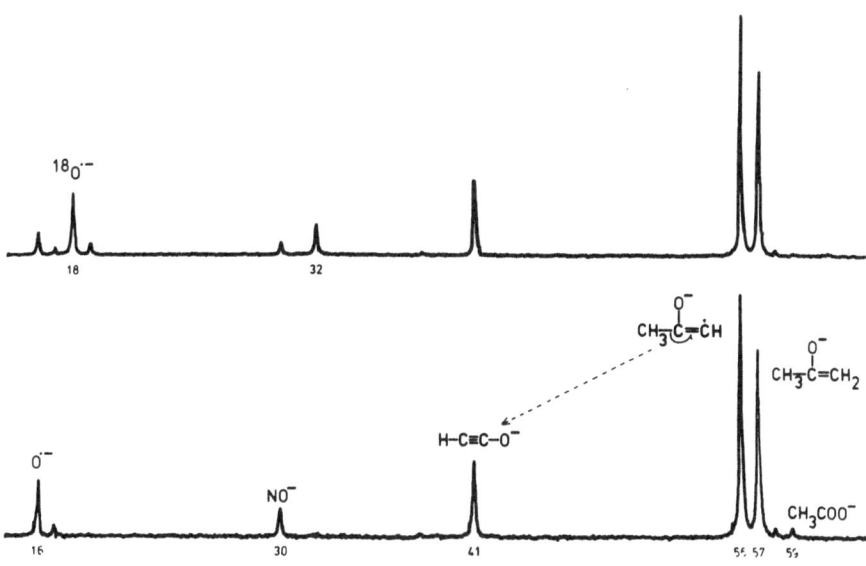

Fig. 2. ICR mass spectra of the N_2O (lower) and $N_2{}^{18}O$ (upper)/$(CH_3)_2CO$
system [2:1 mixtures, total pressure ∿ 3×10^{-5} Torr].

It therefore

$$CH_3-\underset{\underset{\displaystyle O.}{|}}{\overset{\overset{\displaystyle O^-}{|}}{C}}-CH_3$$

I

seemed to us that such an intermediate might also play a role in the reactions which produce $CH_3COCH^{-\bullet}$ and $HCCO^-$ and perhaps even some of the $CH_3COCH_2^-$ ions.

However, as will be seen from Figure 2, when ^{18}O labelled nitrous oxide is used it is only the acetate anion which incorporates the label.

To nullify the suggestion that the oxygen atoms might not actually become equivalent in the intermediate I we have performed similar experiments with dimethylsulphoxide $(CH_3)_2SO$ where electron transfer from oxygen to oxygen through sulphur will doubtedly be a facile process. Here,too, we find no oxygen incorporation into the three major equivalent ions. The double resonance results are the same in this system as for acetone, that is to say:

$$O^{-\bullet} \quad\quad \longrightarrow \quad (M - H_2 - CH_3^{\bullet})^- \quad\quad\quad (5)$$

$$O^{-\bullet} \quad\quad \longrightarrow \quad (M - H_2)^{-\bullet} \quad\quad\quad\quad\quad (6)$$

$$O^{-\bullet}, OH^- \quad \longrightarrow \quad (M - H)^- \quad\quad\quad\quad\quad\quad (7)$$

but as can be seen from a comparison of the acetone and dimethylsulphoxide spectra (Figures 2 and 3) the product distributions are rather different.

Part of the explanation lies in the lower acidity of dimethylsulphoxide in the gas phase[7] but the most striking difference is that almost all the $C_2H_4SO^{-\bullet}$ ions which are formed expel a methyl radical spontaneously to yield $HCSO^-$ as the major ion. This may be accounted for by the lower strength of the CH_3-SO bond of 50 kcal/mole[8] as compared with the CH_3-CO-bond strength of 80 kcal/mole[9].

From all these observations we may conclude that the formation of the (M-H)$^-$ ions takes place through simple H$^+$ abstraction:

$$CH_3COCH_2-H \;+\; OH^- \;\rightarrow\; [CH_3COCH_2^- \overset{+}{\text{---}}H\cdot\text{---}^-OH]^* \;\rightarrow\; CH_3COCH_2^- + H_2O \quad (8)$$
$$(OD^-) \qquad\qquad\qquad\qquad\qquad\qquad\qquad\qquad\qquad (HDO)$$

The reaction of $O^{-\cdot}$ could begin in the same way, but then the radical nature of the ion could lead to a homolytic hydrogen transfer and thus

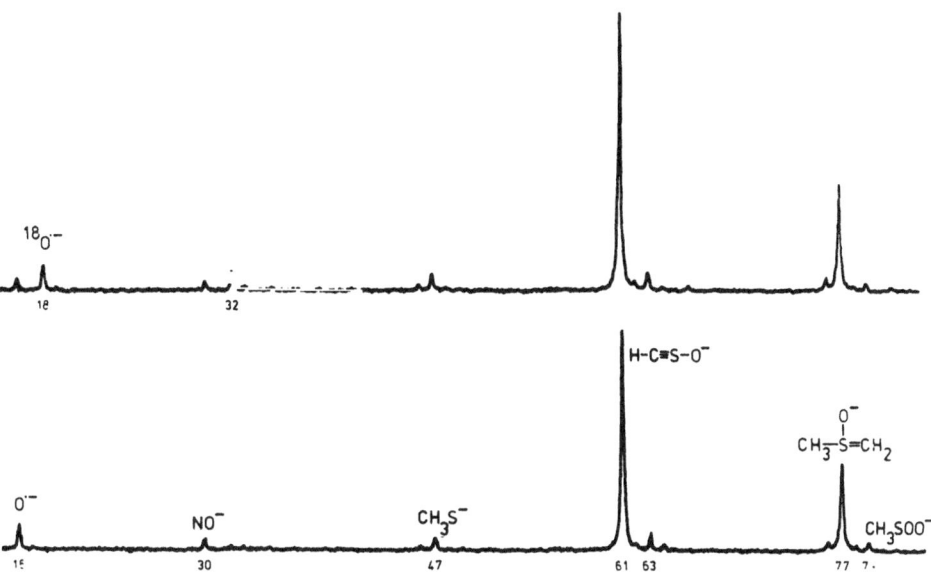

Fig. 3 . ICR mass spectra of the N_2O (lower) and $N_2{}^{18}O$ (upper)/$(CH_3)_2SO$ systems [2:1 mixtures, total pressure $\underset{\sim}{\sim}$ 3×10^{-5} Torr].

water loss:

$$CH_3COCH_2-H \;+\; O^{-\bullet} \;\rightarrow\; \left[\begin{array}{c} O\;\;H \\ \parallel\;\;| \\ CH_3C-\underset{}{C}H----\overset{\cdot\cdot\,O^-}{H+} \end{array} \right]^{*}$$

$$\downarrow$$

$$\left[\begin{array}{c} O^{\bullet} \\ | \\ CH_3C=\underset{}{C}H----\overset{HO^-}{H+} \end{array} \right]^{*} \;\rightarrow\; CH_3COCH^{-\bullet} \;+\; H_2O \qquad (9)$$

We are assuming here of course that Harrison and Jennings[3] are correct in asserting that the structure of the m/e 56 ion derived from acetone is $CH_3COCH^{-\bullet}$ and not $CH_2COCH_2^{-\bullet}$.

Their evidence is strong, but there remains the possibility that some of the non-decomposing ions may have the alternative structure

$$\underset{CH_2-C-CH_2}{\overset{O^-}{|}}$$

which could be formed by a similar mechanism not involving oxygen equivalence:

$$O^{-\bullet} \;+\; CH_3COCH_3 \;\longrightarrow\; \left[\begin{array}{c} H-CH_2 \\ {}^-O \diagup \quad\;\; \diagdown \\ \quad\;\; C=O \\ +H----\underset{}{C}H_2 \end{array} \right]^{*}$$

$$\downarrow$$

$$\left[\begin{array}{c} H\;\;\;CH_2 \\ {}^-O \diagup \quad \diagdown\!\!\!\!\!\diagdown \\ \quad C-O^{\bullet} \\ +H----\underset{}{C}H_2 \end{array} \right]^{*} \;\rightarrow\; H_2O \;+\; \underset{CH_2-C-CH_2}{\overset{O^-}{|}} \qquad (10)$$

[It could be that all the non-decomposing $C_2H_4SO^{-\bullet}$ ions have this alternative structure].

An attempt at the synthesis of CH_3COCD_3 is currently in progress in our laboratory and so it may be hoped that a definitive answer to this question may shortly emerge.

Methyl formate

The principal reactions which occur in the $OH^-/HCOOCH_3$ system have been studied in considerable detail by Riveros, Isolani and coworkers[6d] who report the following results:

$$OH^- + HCOOCH_3 \longrightarrow \begin{cases} HCOO^- + CH_3OH & (11) \\ CH_3O^-.H_2O + CO & (12) \\ CH_3O^- + H_2O + CO & (13) \end{cases}$$

$$CH_3O^- + HCOOCH_3 \longrightarrow CH_3O^-.CH_3OH + CO \qquad (14)$$

where reaction 12 and the analogous reaction 14 exemplify a general feature of the gas phase anionic chemistry of methyl formate.[6d, 10]

Without wishing to intrude too far into their field, we have carried out a brief investigation into the $H_2^{18}O/HCOOCH_3$ system, hoping to confirm the mechanisms which have been proposed for these reactions by Riveros et al.[6d, 10]. The results are however not clearcut.

After allowing for incomplete ($\sim$ 50 %) labelling of the $H_2^{18}O$ sample in the ICR instrument we found that the label content of the formate anions $HCOO^-$ seemed to be about 80 %. A disquieting factor was the fact that this percentage figure was dependent on the water pressure (increasing with increasing pressure). By working with the $H_2^{18}O/HCOOH$ system we were able to show that this was not due to any exchange reaction taking place subsequent to the formation of the formate anion:

$$HCOO^- + H_2^{18}O \xrightarrow{\quad\times\quad} HCO^{18}O^- + H_2O \qquad (15)$$

We then considered the possibility that a hitherto unsuspected reaction
might be occurring:

$$H^- + HCOOCH_3 \longrightarrow CH_4 + HCOO^- \tag{16}$$

for it must always be remembered that it is H^- and not OH^- which is the
primary ion in the water system, though it does react extremely rapidly[11]
to give OH^-:

$$
\begin{array}{c}
e + H_2O \\
\downarrow 6.5 \text{ eV} \\
H^- + H_2O \longrightarrow OH^- + OH^{\bullet}
\end{array}
\tag{17}
$$

The possible direct reaction of H^- can however only satisfactorily be investigated in some dry system and we have therefore studied the $NH_3/HCOOCH_3$
mixture since ammonia gives not only NH_2^- but also a comparable amount of
H^- ions upon electron bombardment ($\sim$ 5 eV). However, we have so far been
unable to eliminate the presence of hydroxide ions from this system and
until that can be done the observation of double resonance signals from
H^- is not conclusive.

The results for the water solvated methoxide anion are also inconclu-
sive because although they appear in the spectra to be 100% labelled
over a water pressure range from 3 to 30×10^{-6} Torr it is also certain
that the water moiety in this ion is in rapid exchange with the neutral
water molecules present in the system. This is made clear by the observation that in a mixture with 16 % deuterated water the cluster ion does
not carry the 16% d_1 content of the reactant hydroxide ions but the
25 % d_1 content of the neutral water molecules.

One result of the $H_2^{18}O$ work is however quite definitive and that is
that the methoxide ion CH_3O^- is completely unlabelled (as is $CH_3O^-.CH_3OH$).

This and the fact that never more than one labelled oxygen is ever in-corporated into the cluster ion $CH_3O^-.H_2O$ shows that it is exclusively the water moiety in the cluster ion which is labelled. This is consistent with the mechanism for reaction 12 sketched out by Riveros et al.[10]:

$$HO^- \ + \ H-\overset{\overset{\displaystyle |}{|}}{\underset{\underset{\displaystyle O}{||}}{C}}OCH_3 \longrightarrow HOH-----^-OCH_3 \ + \ CO \qquad (12a)$$

The observation that the formate ions contain a large percentage of the label is also consistent with the suggestion that these ions are mostly formed through a tetrahedral intermediate

$$^{18}OH^- \ + \ H-\overset{\overset{\displaystyle O}{||}}{C}-OCH_3 \longrightarrow \left[H-\overset{\overset{\displaystyle O^-}{|}}{\underset{\underset{\displaystyle ^{18}O-H}{|}}{C}}-OCH_3 \right]^* \longrightarrow HCO^{18}O^- \ + \ CH_3OH \quad (11a)$$

However, from our results we cannot rule out the possibility that an S_N2 reaction[12,6d] is also occuring to some minor extent:

$$HCOO-CH_3 \ + \ ^{18}OH^- \longrightarrow HCOO^- \ + \ CH_3^{18}OH \qquad (11b)$$

We consider it unlikely that the tetrahedral complex could lose labelled methanol but again this could account for some of the apparent loss of label.

We have also studied the reactions of $O^{-\cdot}$ in the $N_2O/HCOOCH_3$ system and here four further reactions are observed:

$$O^{-\cdot} \ + \ HCOOCH_3 \Big\lbrace \begin{array}{llll} \rightarrow OH^- & + & (M-H)^{\cdot} & (18) \\ \rightarrow OH^{\cdot} & + & (M-H)^- & (19) \\ \rightarrow H_2O & + & (M-H_2)^{-\cdot} & (20) \\ \rightarrow CH_3O^{\cdot} & + & HCOO^- & (21) \end{array}$$

Of course the hydroxide ions formed in reaction 18 react further as outlined above but it is clear from low pressure studies that reaction 21 is
a direct route from $O^{-\bullet}$ to $HCOO^-$ and is in fact the major reaction of the
system whereas the formation of this ion is a less prominent feature of the
$H_2O/HCOOCH_3$ system. Figure 4 records the mass spectra of both the labelled
and unlabelled $O^{-\bullet}$. It will be seen that neither the $(M-H)^-$ nor the $(M-H_2)^{-\bullet}$
ions incorporate any of the label so that they must be formed by direct
H^+ and $H_2^{+\bullet}$ abstraction reactions as outlined in the introduction.

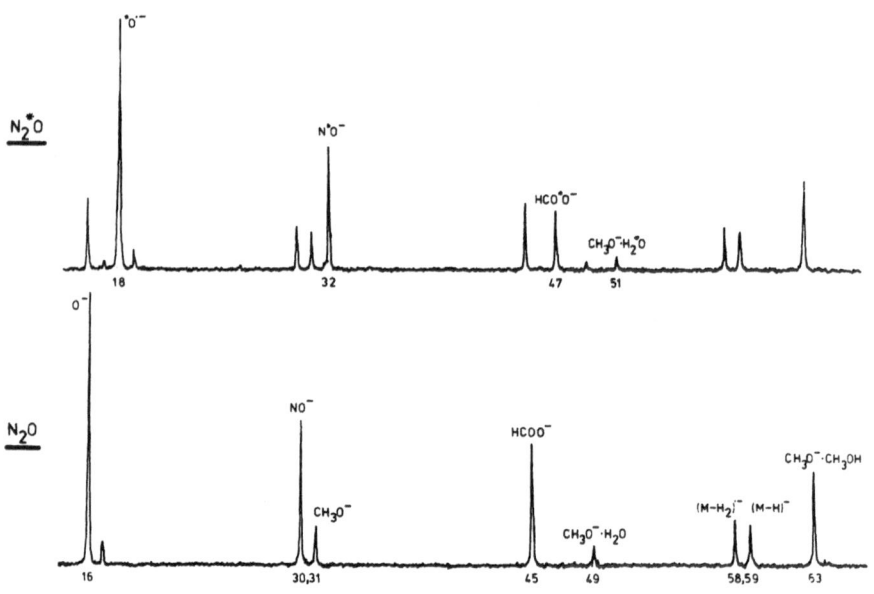

Fig. 4. ICR mass spectra of the N_2O (lower) and $N_2{}^{18}O$ (upper)/$HCOOCH_3$
systems [2:1 mixtures, total pressure $\sim 3 \times 10^{-5}$ Torr].

The degree of label incorporation into the HCOO⁻ ion holds con-
stant at 60 ± 2 % over a range of $HCOOCH_3$ pressures from 1 to 10×10^{-6}
Torr and this can only be explained in terms of two mechanisms: an S_N2
reaction accounting for 40% of the product, and a tetrahedral intermediate
giving the larger percentage of completely labelled product:

$$^{18}O^{-\cdot} \quad + \quad CH_3\text{-}OOCH \quad \longrightarrow \quad CH_3^{18}O^{\cdot} \quad + \quad HCOO^{-} \qquad (21a)$$

$$^{18}O^{-\cdot} \quad + \quad H\text{-}\overset{\overset{O}{\|}}{C}\text{-}OCH_3 \longrightarrow \left[\begin{array}{c} O^{-} \\ | \\ H\text{-}C\text{-}OCH_3 \\ | \\ ^{18}O^{\cdot} \end{array} \right]^{*} \longrightarrow HCO^{18}O^{-} \quad + \quad CH_3O^{\cdot} \quad (21b)$$

Ethylene and propylene oxides

The basic gas phase anionic chemistry of both ethylene oxide $\overline{CH_2CH_2O}$
and propylene oxide $CH_3\overline{CHCH_2O}$ has been carefully established by De Puy et
al.[13] who used a flowing afterglow apparatus to study their reactions with
OH⁻, OD⁻, NH_2^- and CH_3O^-. From our own ICR experiments we are able to con-
cur with their finding that both OH⁻ and OD⁻ react with ethylene oxide to
give one product ion only: m/e 59, $C_2H_3O_2^-$. Although the most appealing
structure for $C_2H_3O_2^-$ is the acetate anion CH_3COO^- it is impossible to
suggest any sensible mechanism by which it could be formed. De Puy et al.[13]
have been able to show that it is the alkoxide anion of glycol aldehyde,
formed in the following manner:

$$OD^{-} \quad + \quad \overset{O}{\underset{CH_2\text{---}CH_2}{\triangle}} \rightarrow \left[\begin{array}{c} D\text{-}H \\ | \\ O\text{-}CH\text{-}CH_2O^{-} \end{array} \right]^{*} \xrightarrow{-HD} O=CH\text{-}CH_2O^{-} \qquad (22)$$

It is not clear why the intermediate should lose HD rather than H_2
(to give DO-CH=CH-O⁻) or HDO (to give CH_2=CH-O⁻).

De Puy et al.[13] have pointed out that the HD loss reaction is endothermic to the extent of 15 kcal/mole[14]-this energy being no doubt supplied
by the strain energy of 28 kcal/mole released in the opening of the three
membered ring.

We have now extended their work to cover the reactions of $O^{-\cdot}$ in the
nitrous oxide/ethylene oxide system (see Figure 5). Four cross product
ions are observed: m/e 17 (OH^-), 41 ($HC\equiv CO^-$), 43 ($CH_2=CH-O^-$) and
59 ($O=CH-CH_2O^-$). Double resonance indicates that as would be expected the
first three (new) ions m/e 17, 41 and 43 are exclusively products of the
reaction of $O^{-\cdot}$.

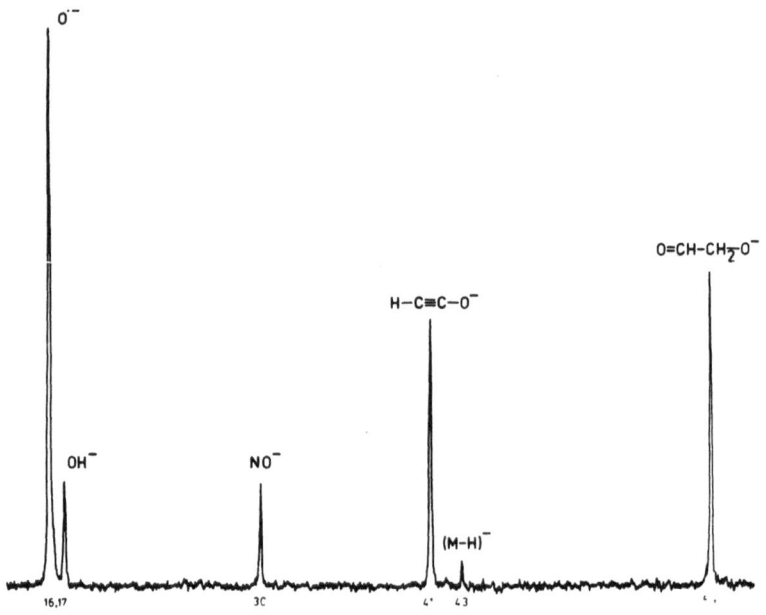

Fig. 5. ICR mass spectrum of the N_2O/CH_2CH_2O system [2:1 mixture,
total pressure $\sim 3\times10^{-5}$ Torr].

Although the double resonance on m/e 59 indicates that $O^{-\cdot}$ is a precursor, a plot of the intensity ratio m/e $59^-/41^-$ tends to zero at zero ethylene oxide pressure indicating that it is formed solely through the intermediacy of the OH^- ions formed in reaction 23.

$$
O^{-\cdot} \ + \ CH_2\!\!-\!\!CH_2 \
\begin{cases}
\rightarrow OH^- \ + \ C_2H_3O^{\cdot} & (23) \\
\rightarrow OH^{\cdot} \ + \ C_2H_3O^- & (24) \\
\rightarrow OH^{\cdot} \ + \ H_2 \ + \ C_2HO^- & (25)
\end{cases}
$$

A change to ^{18}O labelled nitrous oxide brings no change to the ionic products of reactions 24 and 25. This renders unlikely the possibility that either of these ions is formed through any mechanism which involves attack by oxygen upon a corner of the ring leading to a ring openings:

$$
^{18}O^{-\cdot} \ + \ CH_2\!\!-\!\!CH_2 \rightarrow \left[^{18}O^{\cdot}\text{-}CH_2\text{-}CH_2\text{-}O^- \right]^* \longrightarrow \qquad (26)
$$

since one would then expect that the oxygen atoms would very quickly become equivalent. It seems that the first step in both these reactions must be a stripping of H^+ by $O^{-\cdot}$ to give $[C_2H_3O^-]^*$. Perhaps some of these ions remain ring closed and are observed as m/e 43, but the majority (91%) ring open and with the energy thus made available lose a hydrogen molecule:

$$
\left[{}^-CH\!\!-\!\!CH_2 \right]^* \rightarrow \left[\begin{matrix} H \ H \\ | \ | \\ HC\!\!=\!\!C\text{-}O^- \end{matrix} \right]^{**} \xrightarrow{\ -H_2\ } HC\!\!\equiv\!\!C\text{-}O^- \qquad (27)
$$

For this overall scheme for reaction 25 we calculate an exothermicity of $\approx$ 15 kcal/mole.[15]

It is instructive to compare the behaviour of ethylene oxide with
acetaldehyde. Harrison and Jennings[3] have found that in the nitrous oxide/
acetaldehyde system (where there is no ring strain energy to be released)
only the ions OH^-, $C_2H_2O^{-\cdot}$, $C_3H_3O^-$ and CH_3COO^- are formed, there being
no $HC\equiv CO^-$ ion. It is not immediately obvious why the $C_2H_2O^{-\cdot}$ ion (found
by these authors to be mostly $^\cdot CH=CH-O^-$), should be absent in the ethylene oxide system. It could be that the ion, which was found to undergo
collisionally induced electron detachment, is formed from ethylene oxide
but that the energy provided by the ring opening causes it to detach
spontaneously.

So far as its reactions with OH^-, NH_2^- and CH_3O^- were concerned,
De Puy et al.[13] found that propylene oxide reacted only to give an ion
m/e 57, $(M-H)^-$. They drew the clear inference that this must be due to
the abstraction of a methyl hydrogen followed by ring opening to give
the anion of allyl alcohol:

$$B^- \quad + \quad H-CH_2-CH\underset{O}{\overset{\displaystyle\diagup}{\diagdown}}CH_2 \quad \longrightarrow \quad BH \quad + \quad CH_2=CH-CH_2O^- \quad (27a)$$

$$B^- \quad + \quad CH_2=CH-CH_2OH \quad \longleftarrow \quad BH \quad + \quad CH_2=CH-CH_2O^- \quad (27b)$$

We have now been able to measure the relative gas phase acidity of allyl
alcohol (by the method of Brauman and Blair[16]) and we have found that it
lies very close to, perhaps a little below, dimethylsulphoxide. [It was
in fact very difficult to find any double resonance signals in the
$H_2O/CH_2=CH-CH_2OH/(CH_3)_2SO$ system when working under tuning conditions
giving maximum TIC and negligible ion sweep out.] An interesting

consequence arises from the probability that propylene oxide (as a proton

donor only) will have a different acidity from allyl alcohol (the anion

of which it is expected to yield and which will be present only as a pro-

ton acceptor). Depending upon the relative acidities of these two molecules

there will be some acids (BH) for which reaction 27 will either not appear

to go at all or for which it will appear to be exothermic in both directions!

We find that it is the former situation which occurs, acetylene lying in

the dead band:

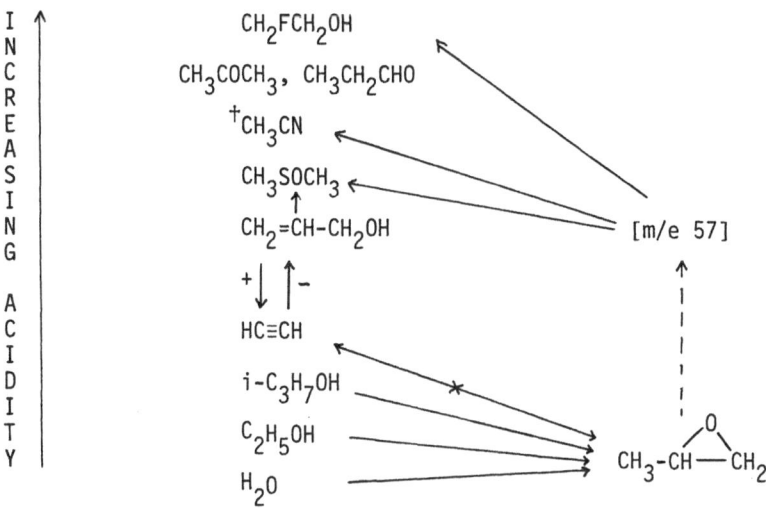

Arrows indicate that the reaction of the anion of the lower

acid has been observed to abstract a proton from the higher acid.

[+]There is some doubt concerning the relative acidities of acetone and

acetonitrile. McMahon and Kebarle[17] working with a pulsed electron beam

high pressure mass spectrometer[18] have reported that at $600^\circ C$ acetonitrile

is the more acidic by a factor of 2.5 kcal/mole. However, we use here the

The actual acidity of propylene oxide (in terms of its ability to donate protons) seems to lie just below that of ethanol since ethoxide anions react giving a small positive double resonance signal.

These results indicate that the (M-H)$^-$ ion derived from propylene oxide has the structure of the alkoxide anion of allyl alcohol since the alternative structures of the anion of acetone and of propanal (which could be formed if OH$^-$ abstracted ring protons) are more acidic than acetonitrile under ICR conditions.

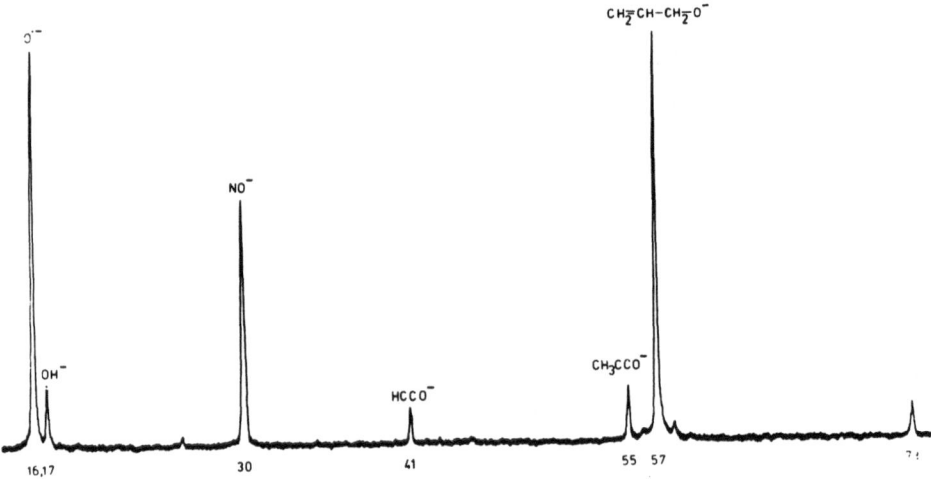

Fig. 6. ICR mass spectrum of the N_2O/CH_3CHCH_2O system (2:1 mixture, total pressure $3x10^{-5}$ Torr).

---

reverse order since Bohme et al.[7] have noted an exothermic proton transfer from acetone to CH_2CN^- in their flowing afterglow apparatus, and since we observe this same order both in our own ICR instrument and in that of Professor K.R. Jennings at the University of Warwick, England, where the gas phase acidity order $(CH_3)_2CO>CH_2FCH_2OH>CH_3CH>to-BuOH$ was recorded [J.H.J. Dawson previously unpublished result].

In the ICR mass spectrum of the nitrous oxide/propylene oxide system we observe m/e 41 and 55 in addition to 17 (OH⁻) and 57 (M-H)⁻ (see Figure 6). There is also a peak at m/e 73 which represents less than 3 % of the total ion intensity. This peak shows double resonance signals from OH⁻ as well as from O⁻· and incorporates ^{18}O. If genuine, it is of course analogous to the m/e 59 ion in the ethylene oxide system, but in view of the fact that it is only of small intensity and was not reported by De Puy et al.[13] we think it best to exclude it from further discussion. The m/e 41 and 57 ions do not incorporate ^{18}O and although observation of the effect upon the peak at m/e 55 is complicated by the presence of the much larger normal peak at m/e 57, it appears that m/e 55 is likewise un-affected. The m/e 41 and 55 ions may be seen as being exactly analogous to the m/e 41 ion of the ethylene oxide system, but with m/e 41 being formed here by the loss of methane rather than a hydrogen molecule:

$$\text{(28a)}$$

$$\text{(28b)}$$

The exothermicity of reaction 28a may be calculated to be about 24 kcal/mole[19] and reaction 28b will also be exothermic to the extent of at least 6 kcal/mole if it is assumed that the heat of formation of $CH_3C{\equiv}CO^-$ (for which data are not available) is less than the heat of formation of $HC{\equiv}CO^-$.

Maleic anhydride

Our interest in the maleic anhydride/nitrous oxide system arose from
the idea that since this organic molecule contains three oxygen atoms there
might be considerable reward from studying its reactions with $^{18}O^{-\cdot}$.

In this respect the system has perhaps not been as interesting as
it might have been, since only one cross product ion and its derivatives
show incorporation of the label. It has however turned out to be a mixture in which each product ion seen may readily be rationalised.

We observe charge transfer, $H^{\cdot}$ (and perhaps H^{+}) transfer and some
more complicated reactions which may all be supposed to commence with a
Michael addition[20] (see for example reaction 32). The ions observed, and
their precursors as determined by double resonance are shown in Table 1.
Six reaction schemes suffice to rationalize all these observations:

$$\longrightarrow OH^{\cdot} \quad + \quad C_4HO_3^{-} \quad \begin{bmatrix} \text{may or may not} \\ \text{occur directly} \end{bmatrix} \qquad (29)$$

$$O^{-\cdot} \quad + \quad C_4H_2O_3 \longrightarrow OH^{-} \quad + \quad C_4HO_3^{\cdot} \qquad (30)$$

$$OH^{-} \quad + \quad C_4H_2O_3 \longrightarrow H_2O \quad + \quad C_4HO_3^{-} \qquad (31)$$

These reactions of $O^{-\cdot}$ probably take place through the first mechanism discussed in the introduction as might the reaction of OH^{-}, but since no
other products are seen to ensue from the collisions of OH^{-} one could postulate a Michael addition[20] followed by a 1,2 elimination of water:

$$(32)$$

Table 1.

	REACTANTS			PRODUCT IONS		
	m/e			m/e	composition	rationalisation
e				16	$O^{-}\cdot$	
	16			17	OH^{-}	
	16	85+		41	$HCCO^{-}$	$(85^{-} - CO_2)$
	16			85	$C_3HO_3^{-}$	$(M + O^{-}\cdot\ -[CHO\cdot])$
	16,	41		95	$C_5H_3O_2^{-}$	$(41^{-} + M - CO_2)$
	16, 17			97	$C_4HO_3^{-}$	$(M - H^{+})$
e, 16+				98	$C_4H_2O_3^{-}\cdot$	$M^{-}\cdot$
	16,		97	107	$C_6H_3O_2^{-}$	$(151^{-} - CO_2)$
	16, 17,		97	151	$C_7H_3O_4^{-}$	$(97^{-} + M - CO_2)$
	16, 17,		97, 107	161	$C_9H_5O_3^{-}$	$(107^{-} + M - CO_2)$

a) The symbol "+" as a postscript denotes a positive double resonance signal.

b) At about 1×10^{-5} Torr $C_4H_2O_3$ and 2×10^{-5} Torr N_2O, m/e 97 and 98 each account for about 40 % of the total product ion concentration, m/e 85 being the next most significant ion.

c) The only changes found in the $N_2^{18}O$ system are that the $C_3HO_3^{-}$ ion shows total incorporation of the label and that $HCCO^{-}$ (which is its daughter by CO_2 loss) incorporates 35% of the label, which we hereafter regard as approximating to $1/3^{rd}$. The double resonance on these ions was:

m/e 85+, 87+   41 and m/e 87+   43.

It could be imagined that the m/e 97 (M-H)$^-$ ions might have a ring opened structure, i.e. O=C=C=CH-COO$^-$, but there is no m/e 53 in the spectra such as would result from the decarboxylation of this ion. Furthermore, the m/e 97 ion seen here is reactive in a way which would not really be expected of a carboxylic acid anion.

Of comparable importance in terms of yield is a pure charge transfer from O$^-$· giving what are, at least formally, molecular anions. Such anions have previously been reported upon by Compton and Cooper alone[21], and in co-authorship with Reinhardt.[22] Working under electron impact conditions in a time of flight mass spectrometer they have measured the electron affinity of maleic anhydride to be 1.3 eV[23] (atomic oxygen = 1.465 eV[24]), and their value for the lifetime of the molecular anion was 250 μs when generated from zero energy electrons. The apparent discrepancy in our results concerning the relative electron affinities may be explained by the fact that the O$^-$· ions formed from N$_2$O are thought to carry an extra 0.38 eV of energy in the form of translational excitation.[4] We have also observed that the molecular anion appears in the absence of N$_2$O in the ICR instrument at near zero eV. We have found it impossible to find a reliable appearance potential maximum because of the extreme difficulty of working below about 1 eV where the trapping plate potential may be more negative than the filament potential. The linewidth of the m/e 98 signal which we have observed does not appear to be abnormally large, but we are not competent to initiate another discussion about the lifetimes of metastable negative ions. The reactions m/e 41 → 95, m/e 97 → 151 and m/e 107 → 161 can all be considered together since they may all be written as:

$$Z^- + C_4H_2O_3 \longrightarrow (Z + C_3H_2O)^- + CO_2 \tag{33}$$

a reaction which may be rationalised as follows:

(33a)

This type of reaction, and proton transfer where relative acidities per-

mit it, will probably summarise the reactions of most non-radical anions

with maleic anhydride.

The reaction m/e 97→107 may now be explained by assuming m/e 97→151 as

the first step. This yields an ion $(C_4HO_3)-\overset{..}{C}H-CH=C=O$ which may undergo

ring opening followed by a second decarboxylation:

(34)

Finally m/e 85 and 41 and their behaviour in the ^{18}O labelled system may

be explained as shown in reaction 35.

Before the results of the $^{18}O^{-}\cdot$ experiments were known it was supposed

that the m/e 85 ion was formed by the loss of a $\cdot$CHO radical rather than

by the consecutive loss of H$\cdot$ and CO. However, the mechanisms which would

allow for the loss of a $\cdot$CHO moiety all predict that it would have con-

tained exclusively the labelled reactant oxygen atom. We believe that

α-cleavage reactions following attack by $O^{-}\cdot$ are very common, even when (as above) it is required that H· rather than a more stable alkyl fragment must be lost.

(35)

Conclusions

The results of the work reported here show that of all the labelled $^{18}O^{-}$· ions which have been generated very few have actually become incorporated into the ionic products of the reactions studied. Thus it might easily be thought that little has been learnt; but that is a view influenced by hindsight, at the outset the expectations were quite different. For a start, it had been suggested that $O^{-}\cdot$ which is not only a very strong gas phase base but also a radical would be such a vigorous reactant that all sence of chemical identity would be lost in its reaction complexes. However, this "Sudden Death" theory is now shown to be completely untennable, for even in the case of maleic anhydride the reactant oxygen atom is found to remain quite distinct in the $[C_4H_2O_4^{-}\cdot]^{*}$ collision complex.

It was a considerable surprise to us to find that in the major reactions of $^{18}O^{-}\cdot$ with acetone no oxygen interchange occurs since this means that the expected tetrahedral intermediate complex is not involved.

The precise nature of the reactions which do not occur is not totally clear, but a comparison with the behaviour of the methyl formate system implies that it is the relatively more acidic hydrogen atoms of acetone which dictate the manner of its reactivity towards $O^{-}\cdot$. For those who would have suggested this in advance we can say that we now provide the proof.

Experimental

ICR mass spectra were recorded on a much modified Varian V5903 instrument fitted with a 2" oil diffusion pump. Sample pressures were measured only approximately on an ionisation gauge placed in a side arm of the main pumping line near to the diffusion pump. The readings quoted are therefore probably too low by a factor of two or three, but the danger of gauge pyrolysis products appearing in the spectra is much reduced. For most of the work reported here the original three section drift cell was used with source drift plate modulation but the instrument is now fitted with a four section (26 x 12 mm) flat cell which has been constructed in our own workshop. In this cell the electron beam is modulated by a control grid, and the stray electrons are ejected in the reaction region only.

$O^{-\cdot}$ anions were generated from nitrous oxide[4] by the dissociative attachment of electrons which were accelerated to 0.6 eV at the source trapping plate.

Except for the critical acidity determining double resonance experiments where the cell was tuned throughout for maximum TIC, it has been normal to tune the source for maximum TIC and subsequent regions for maximum signal strength.

The ^{18}O labelled nitrous oxide used was of 71.5 atom % purity. Although maleic anhydride is a solid it was found to have a sufficiently high vapour pressure to permit it to be introduced through the normal gas/liquid inlet system.

Acknowledgements

Dr. Dawson would like to thank NATO for the granting of a post-doctoral research fellowship (administered by the Science Research Council, London) and we also wish to thank the Netherlands Organisation for Pure Research (SON/ZWO) for the purchase of the basic ICR mass spectrometer. We wish particularly to express our most sincere gratitude to Professor F.S. Klein of the Weizmann Institute, Israel for his most kind gift of the labelled nitrous oxide without which the work reported here could not have been carried out.

References

1. J.H.J. Dawson, Th.A.M. Kaandorp and N.M.M. Nibbering, Org. Mass Spectrom. $\underline{12}$, 330 (1977).

2. D.K. Sen-Sharma, K.R. Jennings and J.H. Beynon, Org. Mass Spectrom. $\underline{11}$, 319 (1976).

3. A.G. Harrison and K.R. Jennings, J. Chem. Soc. Faraday Trans. 2, $\underline{72}$, 1601 (1976).

4. P.J. Chantry, J. Chem. Phys. $\underline{51}$, 3369 and 3380 (1969).

5. (a) M.L. Bender, J. Am. Chem. Soc. $\underline{73}$, 1626 (1951); (b) M.L. Bender, R.D. Ginger and K.C. Kemp, ibid. $\underline{76}$, 3350 (1954); (c) M.L. Bender, R.D. Ginger and J.P. Unik, ibid. $\underline{80}$, 1044 (1958); (d) M.L. Bender, H. Matsui, R.J. Thomas and S.W. Tobey, ibid. $\underline{83}$, 4193 (1961); (e) C.A. Bunton, T.A. Lewis and D.H. Llewellyn, Chem. Ind. (London), 1154 (1954); (f) C.A. Bunton and D.N. Spatcher, J. Chem. Soc. 1079 (1956).

6. See for tetrahedral intermediates proposed in the gas phase: (a) J.H. Bowie and B.D. Williams, Aust. J. Chem. $\underline{27}$, 1923 (1974); (b) J.H. Bowie, ibid. $\underline{28}$, 559 (1975); (c) O.I. Asubiojo, L.K. Blair and J.I. Brauman, J. Am. Chem. Soc. $\underline{97}$, 6685 (1975); (d) J.F.G. Faigle, P.C. Isolani and J.M. Riveros, ibid. $\underline{98}$, 2049 (1976).

7. D.K. Bohme, E. Lee-Ruff and L.B. Young, J. Am. Chem. Soc. $\underline{94}$, 5153 (1972).

8. Calculated by use of ΔH_f° ($CH_3^{\cdot}$) $\underset{\sim}{}$ 33 kcal/mole, ΔH_f° ($CH_3CO^{\cdot}$) = -4 kcal/mole and ΔH_f° (CH_3COCH_3) $\underset{\sim}{}$ -52 kcal/mole; see for these data J.L. Franklin, J.G. Dillard, H.M. Rosenstock, J.T. Herron, K. Draxl

and F.H. Field, Ionization Potentials, Appearance Potentials and Heats of Formation of Gaseous Positive Ions, National Bureau of Standards, Washington D.C., 1969.

9. J.C. Blais, M. Cottin and B. Gitton, J. Chim. Phys. <u>67</u>, 1475 (1970).

10. P.C. Isolani and J.M. Riveros, Chem. Phys. Lett. <u>33</u>, 362 (1975).

11. J.A.D. Stockdale, R.N. Compton and P.W. Reinhardt, Phys. Rev. <u>184</u>, 81 (1969).

12. (a) J.D. Payzant, K. Tanaka, L.D. Betowski and D.K. Bohme, J. Am. Chem. Soc. <u>98</u>, 894 (1976); (b) K. Tanaka, G.I. Mackay, J.D. Payzant and D.K. Bohme, Can. J. Chem. <u>54</u>, 1643 (1976).

13. Veronica M. Bierbaum, C.H. De Pay, R.H. Shapiro and J.H. Stewart, J. Am. Chem. Soc. <u>98</u>, 4229 (1976).

14. S.W. Benson, Thermochemical Kinetics, John Wiley and Sons, Inc., New York, 1969, pp. 48, 201.

15. Calculated by use of ΔH_f° $(O^{-\cdot})$= 24.3 kcal/mole, ΔH_f° $(\overset{O}{\triangle})$ = -12.6 kcal/mole, ΔH_f° $(OH^\cdot)$ = 9.3 kcal/mole, ΔH_f° (H_2) = 0 kcal/mole (see for source ref. 8) and ΔH_f° (C_2HO^-) = -12.9 kcal/mole. The latter value has been taken from J.E. Collin and R. Locht, Int. J. Mass. Spectrom. Ion Phys. <u>3</u>, 465 (1970).

16. J.I. Brauman and L.K. Blair, J. Am. Chem. Soc. <u>92</u>, 5986 (1970).

17. T.B. McMahon and P. Kebarle, J. Am. Chem. Soc. <u>98</u>, 3399 (1976).

18. A.J. Cunningham, J.D. Payzant and P. Kebarle, J. Am. Chem. Soc. <u>94</u>, 7628 (1972).

19. Calculated by use of ΔH_f° $(CH_3-\overset{O}{\triangle})$ = -22 kcal/mole and

ΔH_f^o (CH$_4$) = -17.9 kcal/mole; see for other required data ref. 15 and ref. 8.

20. (a) A. Michael, J. Prakt. Chem. 35, 349 (1887); (b) E.D. Bergman, D. Ginsburg and R. Pappo, Organic Reactions (Ed. R. Adams) 10, 179 1959).

21. C.D. Cooper and R.N. Compton, J. Chem. Phys. 59, 3550 (1973).

22. R.N. Compton, P.W. Reinhardt and C.D. Cooper, Proceedings of the Twentieth Annual Conference on Mass Spectrometry and Allied Topics (American Society for Mass Spectroscopy, June 1972), p. 37.

23. R.N. Compton, P.W. Reinhardt and C.D. Cooper, J. Chem. Phys. 60, 2953 (1974).

24. P.J. Chantry and G.J. Schulz, Phys. Rev. 156, 134 (1967).

Studies in the Chemical Ionization of Hydrocarbons

Raymond Houriet and Tino Gäumann
Department of Physical Chemistry
Ecole Polytechnique Fédérale
CH 1015 Lausanne, Switzerland

1. Introduction

Chemical ionization mass spectrometry(CIMS) has become an important tool in the field of analytical chemistry. CI studies are usually performed using commercial mass spectrometers which are capable of working at relatively high pressures (about 1 torr).

Unfortunately it is difficult to obtain kinetic information from the spectra because of uncertainties in reaction sequence identification. Chemical ionization processes can be conveniently simulated in ICR, since with the longer time scale of ICR a lower pressure and special techniques such as the double resonance can be used. In a series of publications from this laboratory we tried to unravel the fragmentation pattern of hydrocarbons by labelling with deuterium and ^{13}C and using different mass spectrometric techniques such as electron impact and high resolution[1,2,3] field ionization[4], chemical ionization[5] and ICR spectroscopy[6,7]. The use of these techniques allowed us to draw a series of conclusions about unimolecular and bimolecular reaction pathways of hydrocarbon ions. In this publication we try to combine this information with results now obtained from studies in the reactions between methonium and ethyl ions with a number of normal, branched and cycloalkanes.

2. Reactions of CH_5^+ and $C_2H_5^+$ with n-Paraffins

As an example for the study of the reaction mechanisms occuring in the methane chemical ionization of n-paraffins, we have taken n-hexane as a model compound since it presents the following features:

1. n-hexane exhibits four intense ions which have the potential to give information concerning both the intermediate complex and the mechanism of its decomposition

2. it is not such a large molecule that fragmentation occurs by multi-step processes

3. it allows us to make use of the results of our previous studies on alkyl ions to interpret the data [8].

The results of the experiments with n-hexane have been published[6,7] and the most significant conclusions with respect to the different mode of reaction of CH_5^+ and $C_2H_5^+$ will be recalled here.

The ethyl ion reacts as an hydride acceptor with n-hexane to form a $C_6H_{13}^+$ ion as in reaction (1a)

$$C_2H_5^+ \quad + \quad C_6H_{14} \xrightarrow{\quad -C_2H_6 \quad} C_6H_{13}^+ \tag{1a}$$

$$C_6H_{13}^+ \begin{array}{l} \longrightarrow C_2H_4 \quad + \quad C_4H_9^+ \\ \longrightarrow C_3H_6 \quad + \quad C_3H_7^+ \end{array} \tag{1b}$$

The hexyl ion decomposes to form the fragment alkyl ions $C_4H_9^+$ and $C_3H_7^+$ by loss of the olefin molecules C_2H_4 and C_3H_6 respectively. In the course of these decompositions both hydrogen atoms and carbon atoms seem to be equivalent; this indicates that extensive rearrangements must occur in the hexyl ion.

The methonium ion CH_5^+ acts as a proton donor to form an hexonium ion intermediate (reaction (2)).

$$CH_5^+ + C_6H_{14} \xrightarrow{-CH_4} [C_6H_{15}^+]^* \qquad (2a)$$

$$[C_6H_{15}^+]^* \longrightarrow \begin{array}{llll} \rightarrow H_2 & + & C_6H_{13}^+ & (2b) \\ \rightarrow CH_4 & + & C_5H_{11}^+ & (2c) \\ \rightarrow C_2H_6 & + & C_4H_6^+ & (2d) \\ \rightarrow C_3H_8 & + & C_3H_7^+ & (2e) \end{array} \quad \begin{array}{l} \\ \\ -C_2H_4 \end{array}$$

In experiments in which extensive use of labelled reactants was made, we showed that the label retention in the fragments reflects the following major pathways for the decomposition of the $C_6H_{15}^+$ intermediate:

1) loss of H_2 to form a s-$C_6H_{13}^+$ ion

2) loss of a saturated alkane molecule to form a fragment alkyl ion. Unlike the decomposition of the hexyl ion (reaction (1)), this process takes place via a direct carbon-carbon bond scission.

The results of additional experiments carried out with labelled samples of n-pentane and with various unlabelled paraffins support the reaction schemes 1 and 2. In the forthcoming sections, we shall examine in more detail the individual reaction steps encountered in the course of these studies.

2.1. Reaction of $C_2H_5^+$ with n-paraffins

2.1.1. Hydride transfer

The ethyl ion reacts with n-paraffins according to reaction (1). This hydride transfer reaction has been widely recognized as occuring in many ionic systems (for a recent review, see ref. 9). We have observed no re-tention of deuterium in the ionic products when $C_2D_5^+$ is made to react with an n-paraffin molecule; this suggests that the hydride transfer is characterized by the absence of an intermediate complex tight enough to allow

for an H/D process. Lias, Eyler and Ausloos observed adependence of the

reaction rate constant, $k(-H^-)$ upon the sign of the reaction enthalpy ΔH_r.

On the one hand, $C_2H_5^+$ was shown to react with the C_5H_{12} isomers at the

same rate since the reactions to form p-, s- and $t-C_5H_{11}^+$ are all exothermic

(ΔH_r = -32, -111 and -167 kJ/mol respectively). On the other hand, when

the reaction becomes endothermic as in the case of the $s-C_3H_7^+$ reac-

tion, the rate constant of the hydride transfer reaction is seen to de-

crease by a factor of about 30.

In our study, the use of n-paraffins-1,ω-d_6 provides a straightforward

measurement of the relative reactivities of primary and secondary sites

since the transfer of H^- or D^- results in the formation of a secondary or

primary alkyl ion respectively. The results of the hydride transfer to

$C_2H_5^+$ from various samples of labelled n-pentane and n-hexane are pre-

sented in Table 1.

They present three essential features:

a) A slight H/D isotope effect of 1.14 is observed for hexane-1,2,3 -d_7;
    it seems to affect equally the secondary and the primary positions
    (n-pentane-1,5-d_6 and -2,3,4-d_6).

b) A strong positional effect favors the hydride transfer from a
    secondary over a primary hydrogen.

c) The results obtained with d_4-labelled hexanes show that the four
    internal positions in the n-hexane chain seem to be similar with
    respect to the hydride transfer reaction.

The values of the ratio k_s/k_p in Table 1 show that the thermicity of the

hydride transfer reaction plays a role in determining the specifity of this

process. However there is no obvious way for correlating the larger probability to transfer a hydride from a secondary position with the corres-
pondingly larger reaction exothermicity, since there is no constancy of the
ratio k_s/k_p for the few examples listed in Table 1. An increase in this
ratio with increasing chain length can be observed. This could be due
to a possible "screening" of the methylene groups by the terminal methyl
group (steric effect), but as we observe a similar increase of the ratio
k_s/k_p with increasing number of carbon atoms in the methyl substituted
cyclic paraffins, we have to reject this possibility (see section 3.1).

We consider two possible sources of errors that could affect the value
of k_s/k_p. Firstly, it can be argued that the fragment ions might also
contribute to the formation of the hydride transfer product in a series
of higher order specific reactions as exemplified in reaction (3), where
X = H,D. Considering the hydride affinities of the

$$C_2H_5^+ + CD_3(CH_2)_4CD_3 \begin{cases} \xrightarrow{-H^-} prim\text{-}C_6H_8D_5^+ \xrightarrow[-C_3X_6]{} prim\text{-}C_3X_7^+ \xrightarrow[-H^-]{+C_6H_8D_6} sec\text{-}C_6H_7D_6^+ \\ \\ \xrightarrow{} sec\text{-}C_6H_7D_6^+ \xrightarrow[-C_3X_6]{} sec\text{-}C_3X_7^+ \xrightarrow{+C_6H_8D_6} \end{cases} \tag{3}$$

propyl and isopropyl ions [11], the former could transfer a hydride in a
subsequent reaction to form only the secondary hexyl ion thus altering
the significance of the k_s/k_p ratio. Another source that could distort
this ratio might be found in the decomposition of the hexyl ion, e.g. in
a case where the prim-hexyl ion only would fragment to form butyl and propyl fragment ions and the sec-hexyl ion would be stable towards fragmen-

tation. In that case, the measured intensities of the two hexyl ions

($C_6H_8D_5^+$ and $C_6H_7D_6^+$) would not represent the actual relative probabilities

for transferring a hydride or a deuteride from $C_6H_8D_6$:

$$C_2H_5^+ + CD_3(CH_2)_4CD_3 \left[\begin{array}{l} -D^- \rightarrow prim-C_6H_8D_5^+ \longrightarrow C_4X_9^+ \text{ or } C_3X_7^+ \\ \\ -H^- \rightarrow sec-C_6H_7D_6^+ \quad \longrightarrow\!\!\!\times\!\!\!\longrightarrow \end{array} \right. \tag{4}$$

2.1.2. Decomposition of the molecular alkyl ion

When $C_2H_5^+$ reacts with n-paraffins, the molecular alkyl ion thus for-

med loses an olefin molecule to form fragment alkyl ions as shown in

reaction (1'), with (n-m) ≥ 2:

$$C_nH_{2n+1}^+ \longrightarrow C_mH_{2m+1}^+ + C_{(n-m)}H_{2(n-m)} \tag{1'}$$

Apart from this mode of formation, one can also postulate a reaction

in which an alkide entity would be transferred in the same way as a hy-

dride to $C_2H_5^+$. This latter process would result in a direct carbon-car-

bon bond cleavage and thus yield a distinctive distribution of the la-

bel in the product ions.

In Table 2, we report the label retention in the butyl and propyl ions

formed in the reaction of $C_2H_5^+$ with the samples of D and ^{13}C labelled

n-hexane. Both the hydrogen atoms and the carbon atoms are equivalent

in the course of the decomposition of the hexyl ion, similarly to the

behaviour of hexyl ions formed from hexyl halides under electron impact[8].

The rearrangements that allow for a thorough scrambling in the alkyl ions have been discussed in detail in the case of heptyl ions decomposing to butyl ions [2]. In that particular decompositon, it was observed that the probability for a secondary carbon atom to be lost to the C_3H_6 fragment is about 45%, slightly higher than the value corresponding to a complete randomization process (42.9%). On the other hand, the corresponding value for a terminal position is only about 37%. Assuming that the heptyl ion has rearranged to the tert-butyl-dimethylcarbenium ion structure before the fragmentation to tert-butyl ion and propene, the rearrangement of the originally linear heptyl ion was simulated using substituted protonated cyclopropane as intermediate structrure[12]. It seems that this model is suitable for describing the decomposition of the hexyl ion, and thus, also rules out the alkide transfer reaction.

2.2. Reaction of CH_5^+ with paraffins

2.2.1. Proton transfer

The methonium ion reacts with n-paraffins according to the proton transfer reaction (2). The $C_nH_{2n+3}^+$ alkonium ion has not been observed as such but evidence for its formation as an intermediate is provided by the observation of partial retention of the incoming deuteron in the products when CD_5^+ is used as reactant ion. Therefore, the lifetime of this carbonium ion is sufficiently long to allow for a mixing process to occur. The alkonium ion can decompose by losing either H_2 or a saturated alkane molecule and these different steps will be considered in the following sections.

2.2.2. Formation of the molecular alkyl ion

In Table 3, we report the label retention in the parent alkyl ion formed in the reaction of CH_5^+ with n-paraffins. The formation of the alkyl ion $(M-1)^+$ in the reaction of CH_5^+ with n-paraffins is a two-step process, i.e. protonation of the neutral followed by loss of a hydrogen molecule, reaction (5):

$$CH_5^+ + C_nH_{2n+2} \xrightarrow{-CH_4} (C_nH_{2n+3}^+)^* \longrightarrow C_nH_{2n+1}^+ + H_2 \qquad (5)$$

The possibility of a direct hydride transfer to CH_5^+ has been rejected on the following grounds:

- Secondary hydrogens only are involved in the loss of H_2 from protonated hexane-1,6-d_6, whereas $C_2H_5^+$ can accept a hydride both from the primary and secondary positions.

- The hexyl ion formed from CH_5^+ is seen not to decompose [7]. The label retention in the fragment alkyl ions shows that they are formed by the cleavage of one C-C bond in the alkonium ion without rearrangement. On the other hand the hexyl ion formed from $C_2H_5^+$ undergoes substantial rearrangements prior to dissociation (reaction 1b). We thus use this criterion to distinguish between the $C_6H_{13}^+$ products formed from the ethyl and methonium ions. The former is left with a great deal of internal energy as reaction (1a) is exothermic by 109 or 33 kJ/mol depending upon whether a secondary or a primary hexyl ion is formed. While it is true that the enthalpy of reaction (5) is very similar (117 or 41 kJ/mol), there is a possibility that the hydrogen molecule produced may dissipate a large amount of the excess energy of the system. It has been shown in other ion-molecule reaction studies that the H_2 molecule may carry away up to about 85 kJ/mol [13]. Consideration of statistical energy partitioning among the products suggest that most of this excess energy appears as translational energy of H_2 [13].

We have previously observed that secondary hydrogen atoms only were contained in the molecular hydrogen lost from protonated hexane-1,6-d_6. However, results with pentane-2,3,4-d_6 show that there is a slight probability for losing hydrogen atoms from the primary positions. These apparently contradictory facts can be reconciled if we consider the following reaction sequence: the protonation reaction may take place anywhere along the paraffin chain. Two processes are then competing:

a) a process that allows for the mixing of the incoming proton with the hydrogen atoms of the molecule and which results in the retention of the deuterium from CD_5^+.

b) the scission of a C-H or C-C bond, depending upon the location of the protonation.

It seems that process a) allows for about 10 to 15% retention of the incoming hydrogen when the exchange process involves hydrogen atoms in a secondary position, whereas it allows only for about 2 - 4% when primary hydrogen atoms are involved. In the latter case, (protonation on the terminal methyl groups) it seems that the loss of methane (process b)) is a fast reaction as compared to the exchange process a). Moreover, we have observed an important isotope effect in the loss of molecular hydrogen from the hexonium ion, $k(-H_2)/k(-D_2) = 4.5$. A similar isotope effect prevented us from being able to detect the loss of HD in the n-hexane-1,6-d_6

but the analogues loss of H_2 is clearly seen in n-pentane-2,3,4-d_6. The amount of retention of the incoming deuteron from CD_5^+ in the molecular alkyl ion is a most intriguing feature of these results. In order to gain insight into this process, we have used an extensive set of linear and branched saturated hydrocarbons; the results are reported in Table 4. The molecules listed in this table from three distinct classes so far as the retention of D in the molecular alkyl ion is concerned:

1) n-paraffins, with about 12% retention of D.

2) branched paraffins containing at least one methylene group

   (8.5% D retention)

3) branched paraffins containing only primary and tertiary hydrogen

   atoms (2% D retention).

In all three classes of compounds considered above, it is important to notice that the size of the paraffin molecule does not seem to affect the amount of the retention of deuterium. Therefore, this number seems to be directly related to the detailed characteristics of the protonated intermediate, possibly with respect to features such as structure and lifetime. This magnitude of the D retention in the case of the n-paraffins is somewhat unexpected. If we assume that it should be a result of the protonation reaction taking place on a secondary C-H bond, we can calculate the relative probabilities for losing HD or H_2, depending on how many H atoms are equivalent with the incoming D atom. These values are reported in Table 5. A reasonable case would be to consider the equivalency of the incoming D with two hydrogen atoms of a methylene group; in other words this would correspond to a scrambling process restricted to the site of the protonation. The large difference between the calculated retention probability (0.67) and the experimental value (0.88) could possibly be explained by the fact that the exchange equilibrium is not reached in the alkonium ion because of the competing scission process b) considered above. This seems to be quite compatible with the trends observed when the branching in the neutral increases. In that case, a decrease in D retention is observed, while the relative contribution of CH_5^+ to the $(M-1)^+$ ion decreases, as there are less secondary C-H bonds available. We observe that the relative intensity of the product corresponding to the loss of methane from the alkonium increases with branching as previously noted by Clow and Futrell [14].

2.2.3. Formation of the fragment alkyl ions

In Table 6, we report the label retention in the fragment ions formed in the reactions of CH_5^+ (CD_5^+) with n-paraffins. According to reaction (2), the alkonium ion decomposes into fragment alkyl ions by losing a saturated alkane moiety. The loss of CH_4 is seen to involve only the terminal atoms essentially without retention of the incoming hydrogen from the methonium ion (reaction (2c)).

The loss of C_2H_6 is seen to take place by the scission of the C(2) - C(3) bond [7]. This relates directly to the mechanism of the chemical ionization by the methonium ion, i.e. the protonation reaction by CH_5^+ can be visualized as an electrophilic attack occuring at random along the n-paraffin chain, followed by a localized fragmentation step. The label retention in the butyl ion (Table 8) provides a confirmation of this mechanism which was postulated by Field and collaborators [15]. However, we can see that a partial exchange of the deuterium from CD_5^+ takes place when the protonation involves the secondary positions of the alkane chain. As suggested in our previous publication [6], a partial equilibration between the C-H and C-C forms of protonated alkane has to occur prior to fragmentation to butyl ion. The loss of C_3H_8 from $C_6H_{15}^+$ has been shown earlier to proceed via a similar reaction in competition with a process involving two consecutive decomposition steps (see also ref. 7).

3. Reaction of CH_5^+ and $C_2H_5^+$ with cycloparaffins

The distribution of the products formed by the reactions of CH_5^+ and $C_2H_5^+$ with cycloparaffins are reported in Table 7. The ethyl ion is seen to form an unique product by the hydride transfer reaction (6):

$$C_2H_5^+ + c-C_nH_{2n} \longrightarrow C_nH_{2n-1}^+ + C_2H_6 \qquad (6)$$

This reaction has also been observed in a former ICR study on the methane chemical ionization of cyclohexane and a stable cyclohexyl structure has been proposed for the product ion. The methonium ion reacts in a proton transfer reaction (7):

$$CH_5^+ + c-C_nH_{2n} \longrightarrow C_nH_{2n+1}^+ + CH_4 \qquad (7)$$

An interesting question concerns the structure of this stable product ion which could possibly retain the structure of a hydrogen or an olefin molecule, reactions (8) - (9) with (n-m) $\geq$ 2:

$$C_nH_{2n+1}^+ \left\{ \begin{array}{ll} \xrightarrow{-H_2} & C_nH_{2n-1}^+ \qquad (8) \\ \xrightarrow{-C_{(n-m)}H_{2(n-m)}} & C_mH_{2m+1}^+ \qquad (9) \end{array} \right.$$

In addition to these products, CH_5^+ does form a $(M-CH_3)^+$ ion when reacting with methyl substituted cycloparaffins. This suggests that CH_5^+ reacts with the saturated side-chain in a similar way to n-paraffins.

3.1. Reaction of $C_2H_5^+$

The only product formed in the reaction of the ethyl ion with the cycloparaffin is the (M-1)$^+$ ion (reaction (6)). The intensity of this ion is therefore not affected by the possible effects of higher order reactions or decomposition processes as discussed in section 2.1.1. In Table 8 we report the results of the hydride transfer from a series of deuterated cycloparaffins. From the values obtained with the deuterated cyclohexanes, we deduce that there is a slight H/D isotope effect $k(H^-)/k(D^-)$ = 1.19 which is very close to the figure of 1.14 determined in the reaction with n-paraffins. After correction for the isotope effect, an important positional effect is again observed; the ratio k_s/k_p (defined in 2.1.1) has a value of 1.47 for methyl-d_3-cyclopentane and of 2.15 for methyl-d_3-cyclohexane, thus showing a larger value for the larger molecule. We also notice that the hydride transfer from a tertiary position is more probable than from a primary position (methyl-cyclopentane-1d_1 : k_t/k_p = 1.5). It can be concluded that the present experiments do substantiate the values obtained with n-paraffins.

3.2. Reaction of CH_5^+

3.2.1. Proton transfer

The stable product resulting from the addition of a proton to a cycloparaffin has a very small relative intensity (Table 7), in accordance with the ICR results of Clow and Futrell with cyclohexane [14]. These authors noticed a wide variation in the intensity of the $C_6H_{13}^+$ ion whether it is formed in the high-pressure CI (about 1 Torr) conditions or in the ICR. The higher relative intensity in the high-pressure CI was attributed to the high collision frequency that favors collisional deactivation of the $C_6H_{13}{}^+)^*$ species which presumably has vibrational excess internal energy.

3.2.2. Loss of H_2 from $C_nH_{2n+1}{}^+$

The relative losses of H_2, HD and D_2 from protonated cycloparaffins are reported in Table 9. The retention of the incoming hydron from the methonium ion is only 1 to 4%, except for the cycloheptane where it is 9%. No correlation seems to hold between the values of the retention and the molecular parameters of the neutrals. If we refer to the quali- tative model used in section 2.2.2, the ratio of the competing processes varies drastically from one case to an other. An indication that the ex- cess energy in the reactants may play an important role is found in a recent tandem-ICR study of the internal-energy effects in proton trans- fer to ethane by Smith and Futrell[16]. These authors have shown that a series of hydrogen containing ions were able to react as hydride accep- tors or proton donors; the hydride abstraction (or short-lived, unequi- librated proton transfer was seen to decrease relatively to the proton transfer as the internal energy of the reactant ion was decreased by collisional deactivation in the ion source of the tandem instrument. Si- milar considerations might apply to the reaction of $CH_5{}^+$ with cyclopara- ffins but we have not yet carried out any systematic investigations with respect to these effects. The values observed for labelled cyclohexanes (Table 9) correspond to the values which may be calculated by assuming that 4% of the incoming hydron exchanges with D or H and that in the

remaining 96% the hydron is eliminated with a H or D taken at random.
The average deviation amounts to less than 1%. In the case of the me-
thylcycloalkanes, no primary hydrogen atom is lost in the molecular hydrogen as was observed with the n-paraffins (section 2.2.2).

3.2.3. Loss of olefin from $C_nH_{2n+1}^+$

The label retention in the fragment alkyl ions formed in the reaction of CH_5^+ (CD_5^+) with cycloparaffins are reported in Table 10 and 11.
Both the hydrogen atoms and carbon atoms seem to be equivalent in protonated cyclopentane and cyclohexane. However, the relative intensities
of the ions containing the maximum number of labelled atoms are regular-
ly higher than the calculated values. Since we believe that a systematic
error can be excluded, this fact may be taken as evidence for a secondary
isotope effect. However, as the structure of the parent ion is unknown,
we are unable to estimate a numerical value for this effect.

In section 2.1.2, we also observed that the $C_6H_{13}^+$ ion which was formed from the reaction of $C_2H_5^+$ abstracting a hydride from n-hexane showed
a complete randomization of the H and C atoms. We thus conclude that both
$C_6H_{13}^+$ ions have an identical structure. We postulated earlier[7] that the
rearrangements taking place in the hexyl ion prior to decomposition involve the formation of substituted protonated cyclopropyl intermediates
in a similar way to the process of decomposition of the heptyl ion[12].
However, our results do not allow us to decide whether the protonation of
the cyclohexane ring (presumably on a carbon-carbon bond) is followed by
a ring contraction or by the opening of the ring. In an earlier publication on the high-pressure chemical ionization of cycloparaffins, the

ring opening has already been proposed by Field and Munson, on the basic

of the thermochemical data associated with the protonation reaction by

CH_5^+ [15].

The results obtained with methyl substituted cycloparaffins are more

difficult to interpret since different $C_nH_{2n+1}^+$ isomers might be formed

depending upon the site of the protonation. Nevertheless, we notice that

extensive rearrangement occurs in the protonated methylcyclopentane

prior to its fragmentation to butyl and propyl ions. A tendency for the

three methyl hydrogens to remain together is noticeable, and indeed such

a lower participation of primary hydrogens in rearrangements has previous-

ly been reported to occur in many decomposition processes of alkyl ions [17].

This is particularly evident in methyl-d_3-cyclohexane where the unexpected

distribution of the label in the butyl ion (Table 11) matches closely the

results obtained from 7-d_3-heptyl iodide under electron impact [18]. Whether

this indicates that CH_5^+ preferentially protonates the $C_{(1)}-C_{(2)}$ bond in

methylcyclohexane remains an open question, as there is as yet only limited

information on the details of the decomposition of branched alkyl ions. It

is also of interest that in the reaction of CD_5^+ with cycloheptane, the la-

bel distribution in the butyl ion fragment resembles very much the value

obtained in the high pressure CI mass spectrum of n-heptene-1 with D_2O

as the ionizing gas [5], where it was deduced that a secondary heptyl ion

was formed with a D atom in the terminal position. We have recalled the

major features associated with the decomposition of heptyl ions to butyl

ions in section 2.1.2. These results now indicate that the rupture of

the C-C bond of the seven-membered ring takes place at the location of

the protonation by the deuteron, since this process leads to the formation of a heptyl ion labelled in a terminal position.

Conclusions

One of the initial goals of these studies was related to the characterization of the two major reactant ions that are present in the methane chemical ionization techniques. The use of double resonance ICR techniques shows that the methonium and the ethyl ion, each yield distinctive product distributions when reacting with alkanes. The products formed from the ethyl ion do not contribute significantly to our knowledge on the detailed structure of hydrocarbons; we have seen that the fragments are formed in reactions involving numerous rearrangement steps. On the other hand, the products formed from the methonium ion give an insight into many details of the molecular structure of the neutral.

The reaction of CH_5^+ with linear paraffins is of particular interest since the resulting alkonium ion dissociates by simple cleavage of a carbon-carbon bond. This is in contrast to the numerous competing decomposition pathways available to the molecular ions of n-paraffins formed under electron-impact conditions [3]. We may therefore conclude that the methonium ion mass spectra of n-paraffins contain a most useful structural information available from mass spectrometric techniques. Another aspect of these studies is that they contribute to the knowledge of the basic reactions encountered in gas phase studies, e.g. proton transfer and hydride abstraction processes. The fate of the species formed in these primary processes also yields valuable information on the decomposition mechanism of hydrocarbon ions, e.g. alkyl and alkonium ions.

Acknowledgments

The authors wish to acknowledge the participation of Geneviève Boand in the research they have described. We are grateful to the Fonds National Suisse for financial support.

Refences and Notes

1. A. M. Falick and T. Gäumann, Helv. Chim. Acta 59, 987 (1976)

2. D. Stahl and T. Gäumann, "The Mass Spectrum of 1-Heptyl Iodide",

   Org. Mass Spectrom.

3. A. Lavanchy, R. Houriet and T. Gäumann,"The Mass Spectrometric Frag-

   mention of n-Heptane", submitted for publication in the J. Am. Chem.

   Soc.

4. P. Tecon, D. Stahl and T. Gäumann, "Fragmentation of 1-Heptene after

   Field Ionisation", Int. J. Mass Spectrom. Ion Phys.

5. R. Houriet and T. Gäumann, Helv. Chim. Acta 59, 107 (1976)

6. R. Houriet, G. Parisod and T. Gäumann, J. Am. Chem. Soc. 99, 3599

   (1977)

7. R. Houriet and T. Gäumann, "An ICR Study of the Reactions of CH_5^+

   and $C_2H_5^+$ with n-Hexane Labelled with ^{13}C", to be submitted for pu-

   blication in the Int. J. Mass Spectrom. Ion Phys.

8. A. Fiaux, B. Wirz and T. Gäumann, Helv. Chim. Acta 57, 708 (1974)

9. S. G. Lias, R. J. Eyler and P. Ausloos, Int. J. Mass Spectrom. Ion

   Phys. 19, 219 (1976)

10. H_f's taken from a) F. P. Lossing and G. P. Semeluk, Can J. Chem.

    48, 955 (1970); J. L. Franklin, J. G. Dillard, H. M. Rosenstock, J.

    T. Herron, K. Draxl and F. H. Field, Natl. Stand. Ref. Data Ser.,

    Natl. Bur. Stand. No. 26, 1 (1969)

11. The hydride affinity is defined as $H^-A(MH^+) \equiv \Delta H_f(MH^+) + \Delta H_f(H^-) - \Delta H_f(MH_2)$.

    $H^-A(C_3H_7^+) = 1104$ kJ/mole and $H^-A((CH_3)_2CH^+) = 1045$ kJ/mole

12. D. Stahl and T. Gäumann, in Advances in Mass Spectrometry, ed. by

    N. R. Daly, Heyden, London, 1977

13. A. Fiaux, D. L. Smith and J. H. Futrell, J. Am. Chem. Soc. 98, 5773 (1976)

14. R. P. Clow and J. H. Futrell, J. Am. Chem. Soc. 94, 3748 (1972)

15. F. H. Field, M. S. B. Munson and D. A. Becker, Adv. Chem. Ser. No. 58, 167 (1966)

16. R. D. Smith and J. H. Futrell, Int. J. Mass Spectrom. Ion Phys. 20, 347 (1976)

17. G. Parisod and T. Gäumann, see ref. 12

18. A. Fiaux, B. Wirz and T. Gäumann, Helv. Chim. Acta 57, 525 (1974)

Table 1. Hydride Transfer

neutral	label position	product ion	k_s/k_p [c]	ΔH_r (kJ/mole)[d]
$n\text{-}C_6H_{14}$ [a]	$1,6\text{-}d_6$	$C_6H_7D_6^+$	81.5	-109
			3.3	
		$C_6H_8D_5^+$	18.5	-33
	$2,5\text{-}d_4$	$C_6H_9D_4^+$	61.3	
		$C_6H_{10}D_3^+$	38.7	
	$3,4\text{-}d_4$	$C_6H_9D_4^+$	60.9	
		$C_6H_{10}D_3^+$	39.1	
	$1,2,3\text{-}d_7$	$C_6H_6D_7^+$	53.5	
		$C_6H_7D_6^+$	46.5	
$n\text{-}C_5H_{12}$	$1,5\text{-}d_6$	$C_5H_5D_6^+$	63.5	-109
			1.75	
		$C_5H_6D_5^+$	36.5	-33
	$2,3,4\text{-}d_6$	$C_5H_5D_6^+$	39.0	-33
			1.55	
		$C_5H_6D_5^+$	61.0	-109
	$2,4\text{-}d_4$	$C_5H_7D_4^+$	61.0	
		$C_5H_8D_3^+$	39.0	
$n\text{-}C_3H_8$ [b]	$1,3\text{-}d_6$	$C_3HD_6^+$	39.0	-92
			1.9	
		$C_3H_2D_5^+$	61.0	-38

[a] ref.6

[b] ref.9

[c] k_s/k_p = ratio of the rate constants for transferring a hydride from a secondary position over a primary position

[d] ref.10

Table 2.   Label Retention in the Butyl and Propyl Ions Formed in the Reaction of $C_2H_5^+$ with n-Hexane [6,7]

Label position	product ion		stat[a]	product ion		stat[a]
$1,6\text{-}d_6$	$C_4H_3D_6^+$	tr[b]	3.9	$C_3HD_6^+$	tr	0.3
	$C_4H_4D_5^+$	27	25.5	$C_3H_2D_5^+$	3	6.2
	$C_4H_5D_4^+$	47	43.1	$C_3H_3D_4^+$	23	27.7
	$C_4H_6D_3^+$	26	23.1	$C_3H_4D_3^+$	37	40.8
	$C_4H_7D_2^+$	tr	3.9	$C_3H_5D_2^+$	31	21.2
				$C_3H_6D^+$	6	3.6
				$C_3H_7^+$	tr	0.1
$1\text{-}^{13}C$	$^{13}CC_3H_9^+$	67	66.7	$^{13}CC_2H_7^+$	51	50.0
	$C_4H_9^+$	33	33.3	$C_3H_7^+$	49	50.0
$2\text{-}^{13}C$	$^{13}CC_3H_9^+$	65	66.7	$^{13}CC_2H_7^+$	52	50.0
	$C_4H_9^+$	35	33.3	$C_3H_7^+$	48	50.0
$3\text{-}^{13}C$	$^{13}CC_3H_9^+$	67	66.7	$^{13}CC_2H_7^+$	53	50.0
	$C_4H_9^+$	33	33.3	$C_3H_7^+$	47	50.0
$1,6\text{-}^{13}C_2$	$^{13}C_2C_2H_9^+$	37	40.0	$^{13}C_2CH_7^+$	23.5	20.0
	$^{13}CC_3H_9^+$	55	53.3	$^{13}CC_2H_7^+$	54.5	60.0
	$C_4H_9^+$	18	6.7	$C_3H_7^+$	22	20.0

[a] "stat": calculated assuming 20% and 80% hydride transfer from primary, respectively secondary position, to form $C_6X_{13}^+$ (X = H,D)

[b] trace

Table 3. Label Retention in the Parent Alkyl Ion

neutral	label position	product ion	reactant ion CH_5^+	CD_5^+
n-C_6H_{14} [a]	d_0	$C_6H_{12}D^+$		9.7
		$C_6H_{13}^+$	100	90.3
	1,6-d_6	$C_6H_6D_7^+$		10.1
		$C_6H_7D_6^+$	100	89.9
	d_{14}	$C_6D_{13}^+$	97.6	100
		$C_6D_{12}H^+$	2.4	
n-C_5H_{12}	d_0	$C_5H_{10}D^+$		13.5
		$C_5H_{11}^+$	100	86.5
	1,5-d_6	$C_5H_5D_6^+$	99.1	
		$C_5H_6D_5^+$	0.9	
	2,3,4-d_6	$C_5H_5D_6^+$	4.9	
		$C_5H_6D_5^+$	95.1	

[a] ref. 6

Table 4.   Reaction of CD_5^+ with Paraffins. Label Retention in
           the Parent Alkyl Ion

neutral M	% retention of D	relative intensity[a] of $(M-1)^+$	fraction of $(M-1)^+$ formed by CH_5^+
(structure)	10.5	44	.40
(structure)	12	39	.43
(structure)	13.5	34	.48
(structure)	10	51	.54
(structure)	8.5	51	.49
(structure)	8	56	.45
(structure)	8.5	29	.35
(structure)	8.5	b	b
(structure)	2	28	.25
(structure)	2	b	b
(structure)	–	14	–

a
  measured with CH_4 as ionizing gas

b
  not measured, see ref. 14

Table 5.   Expected Probabilities for Losing HD and H_2 from $C_nH_{2n+2}D^+$
           Assuming D to be Equivalent with n Secondary Hydrogen Atoms

n =	1	2	3	4	...	8	meas.
-HD	1.00	0.67	0.50	0.40		0.22	0.88
$-H_2$	0.00	0.33	0.50	0.60		0.78	0.12

Table 6. Label Retention in the Fragment Alkyl Ions Formed
in the Reaction of the Methonium Ion with n-Paraffins

neutral	label position	product ion	reactant ion	
			CH_5^+	CD_5^+
$n\text{-}C_6H_{14}{}^a$	d_0	$C_5H_{10}D^+$		2.5
		$C_5H_{11}{}^+$		97.5
		$C_4H_8D^+$		21
		$C_4H_9{}^+$		79
		$C_3H_6D^+$		17
		$C_3H_7{}^+$		83
	d_{14}	$C_5D_{11}{}^+$	100	
		$C_4D_9{}^+$	87	
		$C_4D_8H^+$	13	
		$C_3D_7{}^+$	85	
		$C_3D_6H^+$	15	
	$1,6\text{-}d_6$	$C_5H_8D_3{}^+$	100	100
		$C_4H_5D_4{}^+$		14
		$C_4H_6D_3{}^+$		86
		$C_3H_2D_5{}^+$		tr
		$C_3H_3D_4{}^+$	tr	12.5
		$C_3H_4D_3{}^+$	51	42
		$C_3H_5D_2{}^+$	33	30.5
		$C_3H_6D^+$	16	15
		$C_3H_7{}^+$	tr	tr

Table 6. (cont'd)

neutral	label position	product ion	reactant ion CH_5^+	CH_5^+
$n\text{-}C_6H_{14}$ [a]	$1\text{-}^{13}C$	$^{13}CC_4H_{11}^+$	51.5	
		$C_5H_{11}^+$	48.5	
		$^{13}CC_3H_9^+$	51.5	
		$C_4H_9^+$	48.5	
		$^{13}CC_2H_7^+$	36	
		$C_3H_7^+$	64	
	$2\text{-}^{13}C$	$^{13}CC_4H_{11}^+$	100	
		$C_5H_{11}^+$	0	
		$^{13}CC_3H_9^+$	53	
		$C_4H_9^+$	47	
		$^{13}CC_2H_7^+$	57.5	
		$C_3H_7^+$	42.5	
	$3\text{-}^{13}C$	$^{13}CC_4H_{11}^+$	100	
		$C_5H_{11}^+$	0	
		$^{13}CC_3H_9^+$	100	
		$C_4H_9^+$	0	
		$^{13}CC_2H_7^+$	45	
		$C_3H_7^+$	55	
	$1,6\text{-}^{13}C_2$	$^{13}C_2C_3H_{11}^+$	0	
		$^{13}CC_4H_{11}^+$	100	

Table 6 (cont'd)

neutral	label position	product ion	reactant ion	
			CH_5^+	CD_5^+
$n\text{-}C_6H_{14}$	$1,6\text{-}^{13}C_2$	$C_5H_{11}^+$	0	
		$^{13}C_2C_2H_9^+$	0.5	
		$^{13}CC_3H_9^+$	98	
		$C_4H_9^+$	1.5	
		$^{13}C_2CH_7^+$	0	
		$^{13}CC_2H_7^+$	71.5	
		$C_3H_7^+$	28.5	
$n\text{-}C_5H_{12}$	$1,5\text{-}d_6$	$C_4H_6D_3^+$	100	
		$C_3H_3D_4^+$	5	
		$C_3H_4D_3^+$	95	
	$2,3,4\text{-}d_6$	$C_4H_3D_6^+$	> 99.5	
		$C_4H_4D_5^+$	tr	
		$C_3H_3D_4^+$	87	
		$C_3H_4D_3^+$	9	
		$C_3H_5D_2^+$	4	

[a] ref. 6,7

Table 7. Product Ions in the Reaction of Cycloparaffins with CH_5^+ and $C_2H_5^+$ [a]

product ion	cyclo-pentane	cyclo-hexane	me-cyclo-pentane	cyclo-heptane	me-cyclo-hexane
$(M+1)^+$	4	2	2	tr	tr
$(M-1)^+$	21(24)	48(36)	28(38)	44(32)	36(32)
$(M+1-CH_4)^+$	-	-	12	-	17
$(M+1-C_2H_4)^+$	51	7	10	-	-
$(M+1-C_3H_6)^+$	-	7	10	24	15

[a] relative intensities are given in parentheses for $C_2H_5^+$ precursor ion

Table 8. Hydride Transfer to $C_2H_5^+$

neutral	label position	loss	meas	stat	k_2/k_p [a]
cyclohexane	d_1	H⁻	92	91.7	
		D⁻	8	8.3	
	$1,1-d_2$	H⁻	85.5	83.3	
		D⁻	14.5	16.7	
	$cis-1,2-d_2$	H⁻	86.5	83.3	
		D⁻	13.5	16.7	
	$1,3-d_4$	H⁻	70	66.7	
		D⁻	30	33.3	
	$1,2,3-d_6$	H⁻	55	50.0	
		D⁻	45	50.0	
	$1,3,5-d_6$	H⁻	53.5	50.0	
		D⁻	46.5	50.0	
me-cyclopentane	$1-d_1$	H⁻	92	91.7	
		D⁻	8	8.3	
	$me-d_3$	H⁻	84	75.0	1.47
		D⁻	16	25.0	
me-cyclohexane	$4-d_1$ [c]	H⁻	83.5	92.9	
		D⁻	16.5	7.1	
	$me-d_1$	H⁻	96	92.9	1.5 [b]
		D⁻	4	7.1	
	$me-d_3$	H⁻	90.5	78.6	2.16
		D⁻	9.5	21.4	

[a] see Table 2   [b] k_t/k_p [c] corrected for 16.5 % isotopic impurity

Table 9. Loss of H_2, HD, D_2 from Protonated Cycloparaffins

neutral	label position	loss:	CH_5^+			CD_5^+		
			H_2	HD	D_2	H_2	HD	D_2
cyclopentane	d_0					1	99	
cyclohexane	d_0					3	97	
	d_{12}			96	4			
	d_1		91.5	8.5		4	89	7
	$1,1\text{-}d_2$		86	14		5	78	17
	$cis\text{-}1,2\text{-}d_2$		86	14		3	79	18
	$trans\text{-}1,2\text{-}d_2$		86	14		5	80	15
	$1,3\text{-}d_4$		67	31	2	3	65	32
	$1,2,3\text{-}d_6$		54	46	tr	2	51	47
	$1,3,5\text{-}d_6$		54	46	tr	2	50	48
me-cyclopentane	d_0					2	98	
	$1\text{-}d_1$		79	21				
	$me\text{-}d_3$		99	1				
me-cyclohexane	d_0					4	96	
	d_{14}			94	6			
	$4\text{-}d_1$		80	20		2	90	8
	$me\text{-}d_1$		100	tr		4	96	tr
	$me\text{-}d_3$		100	tr				
cycloheptane	d_0					9	91	
	d_{14}			90.5	9.5			

Table 10. Reaction of CH_5^+ with Monocycloparaffins. Label Retention in the Fragment Alkyl Ions

neutral	label position	product ion	meas	stat	product ion	meas	stat
cyclopentane	d_o				$C_3H_6D^+$	64	63.6[a]
					$C_3H_7^+$	36	36.4
cyclohexane	d_o	$C_4H_8D^+$	70	69.2[a]	$C_3H_6D^+$	54	53.8[a]
		$C_4H_9^+$	30	30.8	$C_3H_7^+$	46	46.2
	d_{12}	$C_4D_9^+$	32	30.8	$C_3D_7^+$	46	46.2
		$C_4HD_8^+$	68	69.2	$C_3HD_6^+$	54	53.8
	d_1	$C_4H_8D^+$	72	69.2	$C_3H_6D^+$	55	53.8
		$C_4H_9^+$	28	30.8	$C_3H_7^+$	45	46.2
	d_1	$C_4H_7D_2^+$	48	46.2[a]	$C_3H_5D_2^+$	30	26.9[a]
		$C_4H_8D^+$	47	46.2	$C_3H_6D^+$	52	53.8
		$C_4H_9^+$	5	7.7	$C_3H_7^+$	18	19.3
	$1,1\text{-}d_2$	$C_4H_7D_2^+$	49	46.15	$C_3H_5D_2^+$	29	26.9
		$C_4H_8D^+$	44	46.15	$C_3H_6D^+$	52	53.8
		$C_4H_9^+$	7	7.7	$C_3H_7^+$	19	19.3
	$trans\text{-}1,2\text{-}d_2$	$C_4H_7D_2^+$	47	46.15	$C_3H_5D_2^+$	28	26.9
		$C_4H_8D^+$	44	46.15	$C_3H_6D^+$	53	53.8
		$C_4H_9^+$	8	7.7	$C_3H_7^+$	19	19.3
	$cis\text{-}1,2\text{-}d_2$	$C_4H_7D_2^+$	51	46.15	$C_3H_5D_2^+$	28	26.9

Table 10. (cont'd)

neutral	label position	product ion	meas	stat	product ion	meas	stat
cyclohexane	cis-1,2-d$_2$	C$_4$H$_8$D$^+$	44	46.15	C$_3$H$_6$D$^+$	54	53.8
		C$_4$H$_9^+$	5	7.7	C$_3$H$_7^+$	18	19.3
	1,3-d$_4$	C$_4$H$_5$D$_4^+$	21	17.6	C$_3$H$_3$D$_4^+$	6	4.9
		C$_4$H$_6$D$_3^+$	47	47.0	C$_3$H$_4$D$_3^+$	30	29.4
		C$_4$H$_7$D$_2^+$	27	30.2	C$_3$H$_5$D$_2^+$	45	44.0
		C$_4$H$_8$D$^+$	5	5.1	C$_3$H$_6$D$^+$	19	19.6
		C$_4$H$_9^+$		0.1	C$_3$H$_7^+$		2.1
	1,4-d$_4$	C$_4$H$_5$D$_4^+$	24	17.6	C$_3$H$_3$D$_4^+$	11	4.9
		C$_4$H$_5$D$_3^+$	48	47.0	C$_3$H$_4$D$_3^+$	33	29.4
		C$_4$H$_7$D$_2^+$	26	30.2	C$_3$H$_5$D$_2^+$	40	44.0
		C$_4$H$_8$D$^+$	2	5.1	C$_3$H$_6$D$^+$	16	19.6
		C$_4$H$_9^+$		0.1	C$_3$H$_7^+$		2.1
	1,3,4-d$_6$	C$_4$H$_3$D$_6^+$	6	4.9	C$_3$HD$_6^+$	2	0.4
		C$_4$H$_4$D$_5^+$	31	29.4	C$_3$H$_2$D$_5^+$	8	7.3
		C$_4$H$_5$D$_4^+$	44	44.1	C$_3$H$_3$D$_4^+$	31	30.6
		C$_4$H$_6$D$_3^+$	17	19.6	C$_3$H$_4$D$_3^+$	37	40.8
		C$_4$H$_7$D$_2^+$	3	2.1	C$_3$H$_5$D$_2^+$	18	18.4
					C$_3$H$_6$D$^+$	4	2.4
					C$_3$H$_7^+$		0.1

Table 10. (cont'd)

neutral	label position	product ion	meas	stat	product ion	meas	stat
cyclohexane	1,3-$^{13}C_2$	$^{13}C_2C_2H_9^+$	43	40.0	$^{13}C_2CH_7^+$	22	20.0
		$^{13}CC_3H_9^+$	54	53.3	$^{13}CC_2H_7^+$	22	20.0
	1,4-$^{13}C_2$	$C_4H_9^+$	3	6.7	$C_3H_7^+$	16	20.0
		$^{13}C_2C_2H_9^+$	43	40.0	$^{13}C_2CH_7^+$	23	20.0
		$^{13}CC_3H_9^+$	50	53.3	$^{13}CC_2H_7^+$	60	60.0
	1,3,5-$^{13}C_3$	$C_4H_9^+$	7	6.7	$C_3H_7^+$	17	20.0
		$^{13}C_3CH_9^+$	22	20.0	$^{13}C_3H_7^+$	5	5.0
		$^{13}C_2C_2H_9^+$	60	60.0	$^{13}C_2CH_7^+$	45	45.0
		$^{13}CC_3H_9^+$	18	20.0	$^{13}CC_2H_7^+$	42	45.0
cycloheptane	d_o	$C_4H_8D^+$	65.5	60.0[a]			
			34.5	40.0[a]			
	d_{14}	$C_4D_9^+$	35	40.0			
		$C_4HD_8^+$	65	60.0			

[a] values obtained with CD_5^+

Table 11. Reaction of CH_5^+ with Methyl Substituted Cycloparaffins. Label Retention in the Fragment Alkyl Ions

neutral	label position	product ion	meas	stat	product ion	meas	stat
Me-cyclopentane	d_o	$C_4H_8D^+$	71	69.2]a	$C_3H_6D^+$	55	53.8]a
		$C_4H_9^+$	29	30.8]	$C_3H_7^+$	45	46.2]
	$1-d_1$	$C_4H_8D^+$	76	69.2	$C_3H_6D^+$	54	53.8
		$C_4H_9^+$	24	30.8	$C_3H_7^+$	46	46.2
	$1-d_1$	$C_4H_7D_2^+$	52	46.15]a	$C_3H_5D_2^+$	27	26.9]a
		$C_4H_8D^+$	43	46.15	$C_3H_6D^+$	54	53.8
		$C_4H_9^+$	5	7.7]	$C_3H_7^+$	18	19.2]
	$me-d_3$	$C_4H_6D_3^+$	40	29.4	$C_3H_4D_3^+$	19	12.2
		$C_4H_7D_2^+$	42	50.3	$C_3H_5D_2^+$	41	44.1
		$C_4H_8D^+$	16	18.9	$C_3H_6D^+$	35	36.7
		$C_4H_9^+$	2	1.4	$C_3H_7^+$	6	7.0
	$1-^{13}C$	$^{13}CC_3H_9^+$	73	66.7	$^{13}CC_2H_7^+$	49	50.0
		$C_4H_9^+$	27	33.3	$C_3H_7^+$	51	50.0
	$me-^{13}C$	$^{13}CC_3H_9^+$	72	66.7	$^{13}CC_2H_7^+$	51	50.0
Me-cyclohexane	d_o	$C_4H_8D^+$	66	60.0]a			
		$C_4H_9^+$	34	40.0]			
	d_{14}	$C_4D_9^+$	64	60.0			
		$C_4HD_8^+$	36	40.0			

[a] values obtained with CD_5^+

[b] values in brackets from the EI mass spectrum of n-heptyl iodide, 7-d_3 (ref. 18)

Table 11. (cont'd)

neutral	label position	product ion	meas	stat	product ion	meas	stat
Me-cyclohexane	4-d_1	$C_4H_8D^+$	62	60.0	$C_4H_7D_2^+$	39	34.3 [a]
		$C_4H_9^+$	38	40.0	$C_4H_8D^+$	48	51.4
					$C_4H_9^+$	13	14.3
	me-d_1	$C_4H_8D^+$	71	60.0	$C_4H_7D_2^+$	40	34.3 [a]
		$C_4H_9^+$	29	40.0	$C_4H_8D^+$	54	51.4
					$C_4H_9^+$	6	14.3
	me-d_3	$C_4H_6D_3^+$	63 [62] [b]	18.4	$C_4H_5D_4^+$	35	9.2 [a]
		$C_4H_7D_2^+$	10 [91]	47.5	$C_4H_6D_3^+$	33	36.9
		$C_4H_8D^+$	7 [6]	29.7	$C_4H_7D_2^+$	10	39.6
		$C_4H_9^+$	20 [23]	4.4	$C_4H_8D^+$	17	13.2
					$C_4H_9^+$	3	1.1
	1,3-$^{13}C_2$	$^{13}C_2C_2H_9^+$	27	28.6			
		$^{13}CC_3H_9^+$	57	57.1			
		$C_4H_9^+$	16	14.3			
	2,6-$^{13}C_2$	$^{13}C_2C_2H_9^+$	27	28.6			
		$^{13}CC_3H_9^+$	59	57.1			
		$C_4H_9^+$	14	14.3			
	1,3,5-$^{13}C_3$	$^{13}C_3CH_9^+$	11	11.4			
		$^{13}C_2C_2H_9^+$	51	51.4			
		$^{13}CC_3H_9^+$	35	34.3			
		$C_4H_9^+$	3	2.9			

Gas-Phase Polar Cycloaddition Reactions

David H. Russell and M. L. Gross
Department of Chemistry
University of Nebraska-Lincoln
Lincoln NE 68588 U.S.A.

The ion-molecule reaction chemistry of simple alkenes has been extensively studied in many laboratories. At low ion translational energies,
as found in the cell of an ion cyclotron resonance spectrometer, the reactions undoubtedly proceed via an intermediate complex. Although much is
known about the product distributions and rates of reaction of these processes, little is known about the structure(s) of the intermediate. This
is a difficult question because, in many cases, the intermediate cannot
be observed. Its properties can be only inferred by indirect methods such
as studies involving isotopic labels or comparison of fragmentation channels with the mass spectral decomposition of suspected intermediate. Unfortunately, deuterium labelling in simple hydrocarbon systems is not very informative because of extensive isomerizations of hydrogen atoms in the intermediate complex prior to subsequent fragmentation.

Unlike solution chemistry, the short-lived intermediate formed in an
ion-molecule reaction at low pressure (approximately 10^{-5} torr) often will
possess high internal excitation because of the exothermicity of the reaction. Figure 1 shows the potential energy surface for the reaction of the
ethene radical cation with neutral ethene. As can be seen from this profile, a number of structures are energetically feasible for the intermediate complex. In fact, all reasonable isomers of $C_4H_8^{+\cdot}$ may be sampled

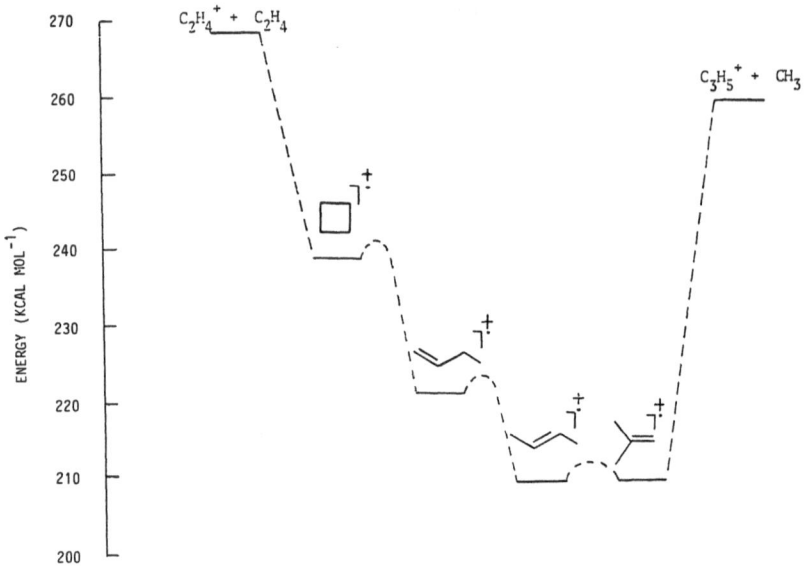

Figure 1. Potential energy diagram for the reaction of the ethene radical cation with neutral ethene. Activation barriers for interconversion of the various C_4H_8 ions are not known. Data taken from H. M. Rosenstock, K. Draxl, B. W. Steiner and J. T. Herron, J. Phys. and Chem. Ref. Data $\underline{6}$, (1977), Supplement No. 1.

prior to decomposition to $C_3H_5^+$ provided there is no large activation ener-

gy for interconversion. The observation of extensive H/D equilibration in

the intermediate is in accord with this concept of the reaction profile.

Less is known about more complicated systems.

Not only is the determination of structure of an intermediate a formi-

dable experimental task, but the actual definition of structure of the

intermediate or reaction complex may be ambiguous. The term "intermediate

complex" should represent those chemical species which are initially for-

med and which result from chemical rearrangements of the primary struc-

ture. Following the initial encounter of an ion and neutral, a number of

events are possible: (1) The complex may retain its original structure

and accommodate any excess excitation in vibrational degrees of freedom.

Sites for localization of charge (i.e., heteroatoms) seem to be effec-

tive in minimizing isomerization which inevitably accompany charge migra-

tion. Fragmentation or decomposition of the complex may then follow.

(2) Prior to decomposition, the intermediate may undergo either hydrogen

atom isomerization or skeletal rearrangements. Study of the complex by

deuterium labelling usually will not be helpful in the former case, but

carbon-13 should be more informative. In any event, the nature of the

initially formed structure may be camouflaged by the rapid isomerization

(3) The excess energy in the complex or in any of its isomerization pro-

ducts may be removed by emission of radiation or collisional stabilization.

The final structure of the complex will depend on the competition between

isomerization and radiative decay or isomerization and collisional stabi-

lization. A thorough understanding of the nature of an intermediate will include its complete history from the moment of initial encounter to the time of fragmentation or stabilization.

The development of organic solution chemistry has benefitted greatly by mechanistic information on intermediate complexes and transition states which serves to correlate available data and provides a basis for predicting new reactions. Perhaps the most important conceptual development in recent years has been selection rules, such as the Woodward-Hoffmann Rules, which are applicable to the broad class of reactions involving molecular rearrangements and cycloadditions. Unfortunately, no analogous "rules" have been uncovered for corresponding gas phase ionic reactions. Accordingly, it is impossible to predict a priori the outcome of reaction involving a specified ion and neutral in the case where the reaction may be analogous to a cycloaddition of the corresponding neutrals.

In this article, we will review the available data on ion-molecule reactions involving reactants which, based on the knowledge of the corresponding neutral/neutral or molecule/molecule chemistry, may proceed via a cycloaddition pathway. Following that, we will discuss some of the studies in our laboratory directed at more complex systems. The discussion constitutes a preliminary attempt to uncover the mechanistic thread which can unify the rich chemistry exhibited in the reactions of ions with the neutral molecules.

Ethene

The ion-molecule chemistry in ethene has been the subject of numerous studies including radiolysis[1], high pressure mass spectrometry[2-4], ion cyclotron resonance (ICR)[4-6] and crossed beam experiments[7]. These investigations have concentrated on the rates of reaction, energy dependence of the reaction cross section, the nature of the collision complex, and the mechanism of secondary ion formation.

The reaction cross section has been reported in several studies[3-5] and has been shown to decrease significantly as the internal energy of the reactive ion is increased from near threshold. Beauchamp, LeBreton, Williamson and Huntress have studied the reaction cross section for $C_3H_5^+$ production by photoionization and found the reaction cross section to decrease by greater than 50% as the energy is increased from near threshold to approximately 5eV above threshold. This observation suggests that as the number of accessible reactant states increases, the probability of reactive collisions decreases, resulting in an overall decrease in the cross section for secondary ion formation.

The radiolysis[1] experiments have provided the most conclusive information concerning the nature of the stabilized intermediate in the ethene ion-molecule reaction. The major products isolated were 1-butene and the cis- and trans-2-butene.

Although the ion-molecule collision complex is not observed without collisional stabilization, its existence can be proved by crossed beam studies.[7] The analysis of product ion angle and translational energy distribution does indicate that at low collision energies (<4eV relative kinetic energies), the reaction proceeds via a long-lived or persistent intermediate. However, at higher energies (>4eV), the product distribution is in accord with a direct mechanism involving a short-lived complex which dissociates before the completion of one full ion-neutral rotation.

Investigations of the $C_3H_5^+$ produced in this reaction have provided the most definitive data concerning the unstable $C_4H_8^{+\cdot}$ intermediate. The results of isotopic distribution in deuterium labelling studies by cross beam experiments[7c] are in good agreement with the ICR[6] and PI[3] studies. However, the more recent data on 1,1-d_2- and 1,2-d_2-ethene by ICR and PI are not in agreement with the "protonated cyclopropane" structure suggested by Wolfgang[7c]. Beauchamp[4] has proposed an intermediate which has the structure of the tetramethylene radical cation. This intermediate can undergo hydrogen rearrangement to one of the butene isomers or fragment to form the initial reactants. This intermediate has been implicated in the fragmentation of cyclobutane[8].

The cycloaddition of neutral ethene is forbidden as a [2+2] concerted reaction according to the Woodward-Hoffmann Rules. It would appear that the corresponding ion-molecule reaction also does not proceed via a concerted [2+2] mechanism. The results are summarized in equation 1. Of course, these observations do not prove the applicability of selection rules to ionic processes.

$$
\text{//} + \text{//} \rightleftharpoons \left(\cdot \quad \rightleftharpoons \quad \cdot \right)
$$

$$
\text{//} + \text{//}^{\uparrow \cdot} \rightleftharpoons \left(\overset{+}{\cdot} \quad \rightleftharpoons \quad \overset{+}{\cdot} \right) \tag{1}
$$

The ion-molecule chemistry of halogenated simple olefins has been investigated, and a four-center-type collision complex system is in good agreement with the observed product distributions.[9] A system which is particularly conclusive is the reaction of perfluoropropene ion with ethene[9h] (equation 2).

$$[CF_3CF=CF_2]^{\ddot{+}} + CH_2=CH_2 \rightarrow [CF_3CF=CH_2]^{\ddot{+}} + CF_2=CH_2 \qquad (2)$$

It is tempting to consider the intermediate as a four-centered or substituted cyclobutane, but the data do not rule out a stepwise mechanism (equation 3) involving an acyclic intermediate which isomerizes via a four-centered transition state. The acyclic intermediate is simply the ionized form of the diradical which is thought to be the intermediate in the corresponding molecule/molecule reaction.

This kind of chemistry may have important analytical applications in locating the double bonds in small quantities of organic materials. For example, the methyl vinyl ether radical cation reacts with various olefins to yield products which are suggestive of either a four-centered intermediate or initial formation of an acyclic intermediate which isomerizes via a cyclobutane-like transition state.[10,11]

Based on the evidence to date, there is no compelling reason to suggest concerted [2+2] ionic cycloadditions of simple olefins. Rather, the intermediate complex appears to be acyclic, and product distributions may be rationalized to occur as ensuing isomerizations which take place via a cyclobutane-like transition state.

Allene

Because of the orthogonal arrangement of the pi-bonds in cumulenes, a [2+2] concerted addition is allowed as a suprafacial, antarafacial process. The chemistry of allene ion plus allene, based on deuterium labelling results, is in accord with a cyclic intermediate[12], but a stepwise process as suggested for the simple olefins is not ruled out. More research is needed to clarify this and related systems.

Butadiene/Ethene

The simplest example for the reaction of a 1,3-diene and an olefin is butadiene and ethene. For molecule/molecule reactions, these reactions combine via an allowed [4+2] cycloaddition to give cyclohexene. This process, known as the Diels-Alder reaction, is extremely useful as a general synthetic method. For ion-molecule reactions, the addition would involve a total of 5 π electrons and could be designated as a [3+2] process if the diene is ionized. For simplicity, we will continue to refer to the reaction as a [4+2] cycloaddition.

In our laboratory, we have investigated the ion-molecule chemistry of 1,3-butadiene with simple olefins, employing ^{2}H and ^{13}C labelling. The results of deuterium labelling are consistent with extensive H/D isomerizations in the collision complex (see Table 1). However, the scrambling reaction is not statistical. This would suggest that (1) the lifetime of the collision complex is too short to allow complete randomization and/or (2) the reaction proceeds via competing reaction mechanisms. The results of carbon-13 labelling are summarized in equations 4 and 5 .

$$CH_2=CH-CH=\overset{*}{C}H_2 \quad + \quad CH_2=DH_2 \quad \xrightarrow[\quad 50\%\quad]{\quad 50\%\quad} \begin{array}{l} m/e\ 68\ +\ CH_3 \\ m/e\ 67\ +\ *CH_3 \end{array} \qquad (4)$$

$$CH_2=CH-CH=CH_2 \quad + \quad \overset{*}{C}H_2=CH_2 \quad \longrightarrow \quad m/e\ 68(100\%)\ +\ CH_3 \qquad (5)$$

If the mechanism is an [4+2] cycloaddition, the intermediate must rapidly open by an allylic cleavage exposing the 3-carbon atom of cyclohexene (originally C-1 in butadiene). Moreover, this ring opening (shown in equation 6) must preempt any double bond isomerization which would ultimately direct ring opening at other sites on the six membered ring. This seems to be an unlikely

$$\overset{*}{\underset{\cdot}{\bigsqcup}} \;+\; \| \;\longrightarrow\; \overset{\cdot}{\underset{*}{\bigcirc}} \;\longrightarrow\; \underset{\cdot}{\bigcirc}^{H} \;\longrightarrow\; C_5H_7^{+} \;+\; \overset{*}{C}H_3 \qquad (6)$$

set of circumstances. Nevertheless, the lifetime of the intermediate $C_6H_{10}^+$ may be sufficiently short that isomerization cannot compete with the rapid ring opening and methyl expulsion. To test this possibility it is necessary to know the chemistry of cyclohexene ion as a function of lifetime or as a function of internal excitation. Field ionization kinetics (FIK) is one experimental approach to obtain this data.

Field ionization kinetics measurements on deuterium labelled cyclohexenes have been interpreted for short-lived ions ($<10^{-10}$s) to indicate fragmentation to produce $C_5H_7^+$ ($M^{+\cdot}$-CH_3) as allylic cleavage from the unrearranged molecular ions followed by loss of the 3-position carbon atom of the cyclohexene.[12] The FIK results may be consistent with a stepwise or concerted formation of cyclohexene in the butadiene/ethene ion-molecule reaction with subsequent loss of methyl occurring exclusively from the terminus of the butadiene moiety. However, this demands that the intermediate has a lifetime of less than 10^{-10} sec. Another small discrepancy in this mechanism is the product distribution in the labelling experiments on 3,3,6,-d_4-cyclohexene (FIK) and the ion-molecule reaction product ions of 1,1,4,4,-d_4-butadiene (ICR). The [4+2] cycloaddition should produce the 3,3,6,6,-d_4-cyclohexene in this latter case. We observe approximately 5% loss of CD_3 whereas the FIK data report no loss of CD_3 even in the long-lived ions (approximately 10^{-6}). Obviously, an important experiment needed to complement the present data is an FIK study of cyclohexene-3^{13}C.

Without further experimental data, we feel that the requirements of the [4+2] cycloaddition are too severe. An acylic ion-molecule collision complex can be employed to explain the data (see equation 7), and this mechanism does

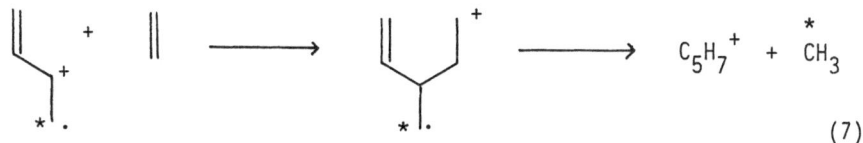

$$\text{(7)}$$

not possess the stringent requirements of the [4+2] process. Loss of me-

thyl occurs exclusively from the single branch point. In the case of buta-

diene-1-^{13}C, the branched methyl will contain the C-13 label in 50% of the

intermediates. Favored cleavage at the branch point will give the observed

1:1 loss of C-12 and C-13. Moreover, the preference for the methyl loss

occurring from the butadiene moiety is explained.

In addition, it should be noted that a [3+2] cycloaddition (either con-

certed or stepwise) cannot be ruled out (equation 8). Further experiments

Table 1. Loss of Methyl from the Intermediate Complex in Reaction of

1,3-Butadiene with Ethene

Butadiene	Ethene	CD_3	CD_2H	CDH_2	CH_3
C_4D_6	C_2H_4	34(17)^c	40(50)^c	24(30)^c	2(3)^c
C_4H_6	C_2D_4	3(3)	24(30)	45(50)	28(17)
1,1,4,4-d$_4$	C_2H_4	10(3)	38(30)	39(50)	13(17)

a) Reprinted from M. L. Gross, D. H. Russell, R. Phongbetchara and
   P. H. Lin, Adv. in Mass Spectrometry, Vol. 7, in press
b) Observed at 12 eV of ionizing energy. All intensities have been
   corrected for ^{13}C.
c) Values in parentheses are calculated for complete scrambling.

to test this mechanism are in progress.

$$\text{(8)}$$

The selectivity of the ion-molecule chemistry of ionized 1,3-butadiene with simple olefins has been reported previously by Gross and Lin.[13] In this study, the ability of ion-molecule reactions to distinguish isomeric neutral species is illustrated. For example, the dominant secondary ion products of the system 1,3-butadiene/1-butene and 1,3-butadiene/2-butene (shown in equation 9 and 10) are characteristic of the neutral structure. For 1-butene, the favored cleavage is loss of ethyl which is the group attached to the double-bonded carbon. Similarly, the complex formed with 2-butene undergoes facile loss of methyl. In similar manner, it was possible to identify the isomeric pentenes. The mechanism for these reactions is not yet established, but a

$$C_4H_6^{+\cdot} + 1\text{-}C_4H_8 \rightarrow [C_8H_{14}^{+\cdot}]^* \rightarrow C_6H_9^+ + C_2H_5^\cdot \qquad (9)$$

$$C_4H_6^{+\cdot} + 2\text{-}C_4H_8 \rightarrow [C_8H_{14}^{+\cdot}]^* \rightarrow C_7H_{11}^+ + CH_3^\cdot \qquad (10)$$

[4+2] cycloadduct or an acyclic intermediate can fit the data.

In related studies, Nibbering, van Doorn, Ferrer-Correia, and Jennings have studied the ion-molecule chemistry of 1,3-butadiene/methyl vinyl ether by utilizing isotopically labelled reactants and by comparison with the mass spectrum of the proposed intermediate.[14] These results are consistent with the formation of the 4-methoxycyclohexene via [4+2] cycloaddition (equation 11). Since the ion chemistry of acyclic intermediates is not yet established, it is not possible to rule out conclusively another mechanism.

$$\overset{\cdot+}{\underset{}{\big]}} \quad + \quad \overset{OCH_3}{\big\|} \quad \longrightarrow \quad \overset{\cdot+}{\left[\bigcirc - \overset{OCH_3}{}\right]} \qquad (11)$$

Furan/Butadiene

In a similar study, we have investigated the ion-molecule chemistry of ionized furan with neutral 1,3-butadiene. The reaction proceeds through a detectable intermediate complex (m/e 122) which undergoes loss of a methyl radical (equation 12). The results of this system are consistent with a [4+2] cycloadduct collision complex. Deuterium labelling experiments may be interpreted

$$C_4H_4O^{\overset{+}{\cdot}} + C_4H_6 \rightarrow [C_8H_{10}O]^{\overset{+}{\cdot}} \rightarrow C_7H_7O^+ + CH_3^{\cdot} \qquad (12)$$

as the methyl loss arising from the original butadiene molecule after transfer of a hydrogen atom from the 2-position of the original furan molecule. The results of C-13 labelling suggest that the carbon atom lost arises from the terminal position of the butadiene (approximately 80%) as well as the two central carbon atoms (approximately 20%).

The mechanism which accounts for these results is shown in Scheme 1. As can be seen, the intermediate is a [4+2] cycloadduct, and the furan radical cation serves as the dienophile. Pathway (1) shows the major loss of methyl which is initiated by an energetically favored ring opening

Scheme 1

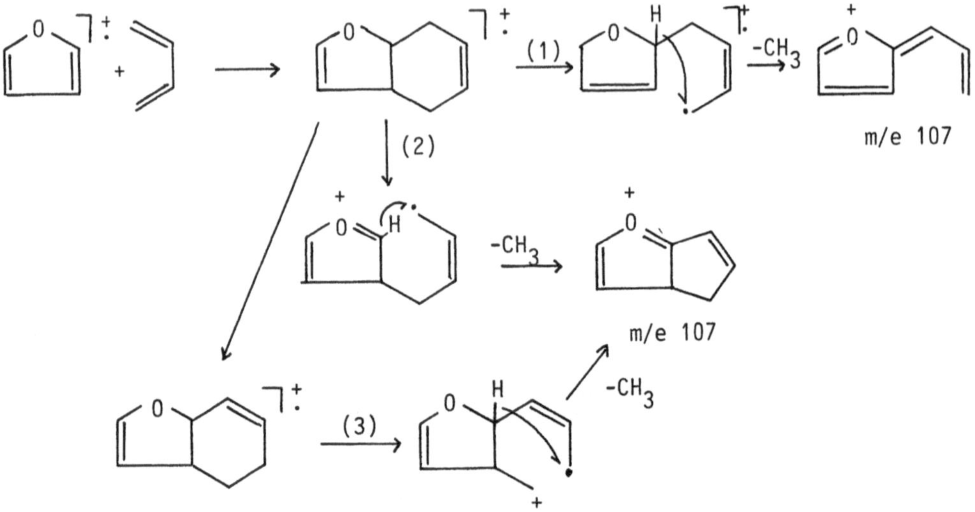

m/e 107

m/e 107

exposing the terminal carbon of the original butadiene and placing the

positive charge on the oxygen atom. To account for the major loss of

CH_2D (approximately 76%) when 2,5-d_2-furan is used as the reactant, the

rearranging hydrogen is shown to migrate from the 2-position of the furan

moiety. Two facile ring openings (pathways 1 and 2) are possible, and

both are directed by the double bond in the six-membered ring and by the

oxygen atom. Some undetermined ratio of these account for approximately

80% of the methyl loss. A third, less facile, ring opening (pathway 3)

occurs after double bond isomerization in the six-membered ring and ex-

poses the center carbon atoms of the original butadiene molecule. The

structures of the m/e 107 ions are speculative; nevertheless, the six-

membered ring must form in at least 20% of the intermediate complexes

in order to lay open the center butadiene carbons as sites for methyl

loss.

It may be pertinent to note the similarity in the ion chemistry of furan and methyl vinyl ether with butadiene. The furan ion may be regarded as vinyl ether analog rather than an aromatic radical cation.

o-Quinodimethane/Styrene

The stereochemical configuration of either the reacting ion or the neutral may play an important role in the observed ion-molecule chemistry. In the previous studies involving neutral butadiene, especially in the furan/butadiene system, the butadiene is present primarily as the trans-isomer. However, the configuration of ionized butadiene is not known. To explore the chemistry of cis-diene radical cations, one of the systems we have selected is the o-quinodimethane radical cation.[15] This radical cation may be generated in a number of ways as illustrated by equation 13.

$$(13)$$

The o-quinodimethane radical cation reacts readily with neutral 1,3-butadiene and with styrene. The o-quinodimethane/styrene ion-molecule chemistry has been studied extensively by deuterium labelling of reactants and by comparisons with the unimolecular fragmentation of the suspected intermediates and has been reported previously.[13] Briefly, the dominant ion-molecule products observed are the collision complex ($C_{16}H_{16}^{+\cdot}$;m/e 208) and loss of C_6H_6 from the complex ($C_{10}H_{10}^{+\cdot}$; m/e 130). The results of deuterium labelling experiments indicate that the loss of neutral benzene from the collision complex proceeds in a highly specific way.

If this system involves a [4+2] cycloaddition mechanism, the ion-molecule

collision complex would possess initially the 2-phenyltetralin structure

(see equation 14). Since the deuterium labelling experiments on the ra-

dical cation

$$\longrightarrow C_{10}H_{10}^{+} + C_6H_6 \qquad (14)$$

and the reacting neutral are indicative of partial H/D equilibration, the

analogous deuterium labelled 2-phenyltetralins were studied by normal mass

spectrometry and defocussed metastable studies. These studies were inter-

preted   as a 1,3-elimination of C_6H_6 from the molecular ion of 2-phenyl-

tetralin.   Likewise, the elimination of C_6H_6 from the ion-molecule inter-

mediate may be rationalized as occurring in an analogous manner. It should

be noted that this evidence alone is highly suggestive but not proof of

a cycloadduct, only that the collision-complex in the ion-molecule reac-

tion and 2-phenyltetralin decompose via a common intermediate.

Styrene/Styrene

The ion-molecule chemistry of ionized styrene with neutral styrene

is similar to the chemistry of o-quinodimethane/styrene in that the same

secondary ions are observed.[14,15] Assuming the mechanism is [4+2] cyclo-

addition, there are two feasible structures for the intermediate, namely,

1-phenyltetralin and 2-phenyltetralin (see equations 15 and 16). It is

$$\text{(15)}$$

$$\text{(16)}$$

possible also that the proposed intermediates (I and II) do not isome-

rize as shown due to the forbidden nature of the [1,3]-hydrogen shift which

is required for their direct interconversion. The results of deuterium la-

belling on the α- and β-positions of styrene are interpreted to indicate

a 1-phenyltetralin intermediate. To resolve this question, a variety of

deuterium labelled 1-phenyltetralins are being studied to update the ori-

ginal publication.

$\underline{C_6H_6}$

In our laboratory, we have a long standing interest in the bimolecular

chemistry of C_6H_6 radical cations. The results of these studies on benzene

and its acyclic isomers have been reported.[18] To complement the previous

study and to explore further the possibility of the ionic cycloaddition

reaction, we have begun a study of the ion-molecule chemistry of 3,4-di-

methylene cyclobutene (III) and fulvene (IV). These molecules may be

viewed as both a cis and trans diene, respectively. Fulvene, for example,

has been found to act as both a diene and a dienophile in molecule/mole-
cule reactions (Diels-Alder chemistry).

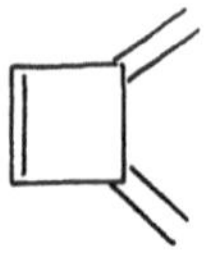

(III)

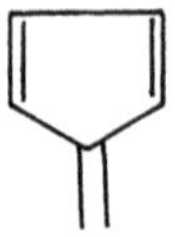

(IV)

We have found that these two cyclic benzene isomers react readily with
neutral 1,3-butadiene to produce a variety of secondary ions (see equation 17).

$$C_6H_6^{+\cdot} + C_4H_6 \quad \begin{array}{l} \nearrow \ C_{10}H_{11}^+ \ + \ H^\cdot \\[1em] \rightarrow \ C_9H_9^+ \ + \ CH_3^\cdot \\[1em] \rightarrow \ C_8H_8^{+\cdot} \ + \ C_2H_4 \\[1em] \searrow \ C_7H_7^+ \ + \ C_3H_5^\cdot \end{array} \qquad (17)$$

To explore the mechanism of this reaction, the isotopic distribution
of H/D and carbon-13/carbon-12 have been utilized. The results of deuterium
labelling are consistent with extensive H/D randomization. However, this
randomization of H/D is not statistical and may be interpreted as a
preference for loss of the original butadiene hydrogens.

The results of carbon-13 labelling are also consistent with the pre-
ferred loss of the original butadiene carbon atoms. In the present study,
the carbon-13 labelling experiments were performed using 1-^{13}C-1,3-buta-

diene. We are presently preparing $2,3-^{13}C_2-1,3$-butadiene in order to explore further the mechanism of the reactions.

Allyl Ion Chemistry

Cycloadditions of closed shell ions (even electron ions) in solution have been investigated extensively and the results are summarized in three recent reviews.[19] This area of polar cycloadditions is ideally suited for integration of our knowledge of gas-phase and solution reactivity. Unfortunately little is known about the gas-phase chemistry.

The simplest example of an allowed cycloaddition involving a total of 6π-electrons and a closed shell reactant ion is the reaction of the allyl ion with 1,3-butadiene (equation 18). In the gas-phase, $C_3H_5^+$ does react with butadiene

$$\text{(18)}$$

to form $C_6H_8^{+\cdot}$ and $C_5H_7^+$ as shown in equation (19). As yet, the mechanism for this process is unknown. Nevertheless, it can be anticipated that systems of this nature will exhibit extensive chemistry.

$$C_3H_5^+ + C_4H_6 \longrightarrow [C_7H_{11}^+]^* \left\langle \begin{array}{l} C_6H_8^{+\cdot} + CH_3^{\cdot} \\ \\ C_5H_7^+ + C_2H_4 \end{array} \right. \tag{19}$$

Conclusion

It is very clear that investigations of ion-molecule reactions based on analogous molecule/molecule processes which proceed as cycloadditions

228

will show rich ion-chemistry. For systems modelled after [2+2] cycloadditions, there appears no compelling evidence in favour of a [2+2] concerted addition for thermal olefin ions. Rather, the results can be rationalized if an acyclic intermediate is invoked. Thermal [2+2] cycloadditions are forbidden and proceed via a diradical intermediate. For the
ion chemistry, the ionized diradical is suggested as the intermediate
which can isomerize by hydrogen rearrangements or undergo skeletal rearrangements via a cyclobutane-like transition state.

The prototype [4+2] ionic cycloaddition is the reaction of the butadiene ion with ethene. Although a thermally allowed molecule/molecule reaction, the ion chemistry is readily explained by an acyclic intermediate.
It is possible that ionized butadiene should be represented as V, and the
chemistry exhibited

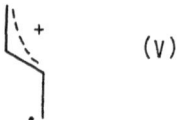   (V)

by this species is determined by preferential attack of the positively
charged site on the electron-rich olefin. In the case of butadiene and
ethene, the loss of methyl takes place by H-transfer in the acyclic intermediate followed by C-C bond scission. When furan or methyl vinyl ether
are substituted for ethene, the acyclic intermediate may cyclize to give
a [4+2] adduct or the initial reaction occurs by a concerted [4+2] process. The presence of the oxygen heteroatom in the intermediate complex
appears to minimize H/D interchanges by providing a site for charge localization. Moreover, with the charge localized on oxygen, as in the furan
ion, the cycloaddition may be a true [4+2] process involving 6π electrons.

Ionic dienes, which are forced cisoid such as the o-quinodimethane radical cation or the styrene ion, appear to react via a [4+2] cycloaddition. These observations suggest that stereochemical effects may be important in determining the course of these extremely rapid reactions.

Although many of these initial studies of ionic cycloadditions have made use of radical cation reagents, closed shell systems, such as the allyl ion, exhibit rich chemistry in situations where a [4+2] cycloaddition may pertain. Studies of these systems should contribute to a better understanding of polar cycloadditions in solution.

It is our hope that this brief review will stimulate other research in the area of gas-phase polar cycloadditon reactions. It is certain that the chemistry exhibited by these systems will be rich and varied. Determination of the mechanisms of the reactions will constitute an important contribution to the fundamental understanding of reaction dynamics.

Acknowledgement

The generous support of the National Science Foundation (Grant CHE76-23549) is gratefully acknowledged.

References

1. G. G. Meisels, J. Chem. Phys. 42, 2328 (1965) and 42, 3237 (1965)

2. F. H. Field, J. L. Franklin and F. W. Lampe, J. Am. Chem. Soc.
   79, 2419 (1957); F. H. Field, ibid. 83, 1523 (1961); T. O. Tiernan and
   J. H. Futrell, J. Phys. Chem. 72, 3080 (1968); J. J. Myher and A. G.
   Harrison, Can. J. Chem. 46, 101 (1968); S. Wexler and R. Marshall, J.
   Am. Chem. Soc. 86, 781 (1964)

3. L. W. Sieck and P. Ausloos, J. Res. Natl. Bur. Stand. A76, 253 (1972)

4. P. R. LeBreton, A. D. Williamson and J. L. Beauchamp, J. Chem. Phys.
   62, 1623 (1975)

5. M. L. Gross and J. Norbeck, ibid. 54, 3651 (1972)

6. M. T. Bowers, D. D. Elleman and J. L. Beauchamp, J. Phys. Chem. 72,
   3599 (1968)

7. (a) Z. Herman, A. Lee and R. Wolfgang, J. Chem. Phys. 51, 452 (1969)

   (b) A. Lee, R. L. Leroy, Z. Herman and R. Wolfgang, Chem. Phys. Lett.
       12, 569 (1972)

   (c) J. Werner, A. Lee and R. Wolfgang, ibid. 13, 613 (1972)

8. C. Lifshitz and T. O. Tiernan, J. Chem. Phys. 55, 3555 (1971)

9. (a) R. M. O'Malley and K. R. Jennings, Int. J. Mass Spectrom. Ion
       Phys. 2, 441 (1969)

   (b) R. M. O'Malley and K. R. Jennings, ibid. 11, 89 (1973)

   (c) V. G. Anicich, M. T. Bowers, R. M. O'Malley and K. R. Jennings,
       ibid. 11, 99 (197

   (d) A. J. Ferrer-Correia and K. R. Jennings, ibid. 11, 111 (1973)

   (e) V. G. Anicich and M. T. Bowers, ibid. 12, 231 (1973)

   (f) V. G. Anicich and M. T. Bowers, J. Am. Chem. Soc. 96, 1279 (1974)

9. (g) C. J. Drewery, G. C. Goode and K. R. Jennings, Int. J. Mass Spectrom. Ion Phys. $\underline{20}$, 403 (1976)

   (h) C. J. Drewery, G. C. Goode and K. R. Jennings, ibid. $\underline{22}$, 211 (1976)

10. C. J. Drewery and K. R. Jennings, ibid. $\underline{19}$, 287 (1976)

11. (a) J. J. Myher and A. G. Harrison, J. Phys. Chem. $\underline{72}$, 1905 (1968)

    (b) M. T. Bowers, D. D. Elleman, R. M. O'Malley and K. R. Jennings, ibid. $\underline{74}$, 2583 (1970)

12. P. J. Derrick, A. M. Falick and A. L. Burlingame, J. Am. Chem. Soc. $\underline{94}$, 6794 (1972)

13. M. L. Gross, P. H. Lin and S. J. Franklin, Anal. Chem. $\underline{44}$, 974 (1972) P. H. Lin, Ph. D. Thesis, Univ. of Nebraska, 1973

14. R. Van Doorn, N. M. M. Nibbering, A. J. V. Ferrer-Correia and K. R. Jennings, private communication

15. P. H. Lin, Ph. D. Thesis, University of Nebraska, 1973 M. L. Gross and P. H. Lin, Am. Soc. Mass Spectrom. 21st Annual Conference on Mass Spectrom. and Allied Topics, San Francisco, Ca, 1973

16. C. L. Wilkins and M. L. Gross, J. Am. Chem. Soc. $\underline{93}$, 895 (1971)

17. M. L. Gross, C. L. Wilkens and T. G. Regulski, Org. Mass Spectrom. $\underline{5}$, 99 (1971)

18. D. H. Russell and M. L. Gross, Am. Soc. of Mass Spectrom., 25th Annual Conference on Mass Spectrom. and Allied Topics, Washington D.C., 1977

19. H. M. R. Hoffmann, Angew. Chemie, Int. Edit. $\underline{12}$, 819 (1973); R. R. Schmidt, ibid. $\underline{12}$, 212 (1973); C. K. Bradsher, Adv. in Heterocyclic Chem. $\underline{16}$, 289 (1974)

An Ion Cyclotron Resonance Study

of an Organic Reaction Mechanism

T. B. McMahon
Department of Chemistry
University of New Brunswick
P. O. Box 4400, Fredericton, New Brunswick, Canada E3B 5A3

Introduction

The detailed mechanisms of organic reactions in solution have pre-
occupied physical chemists for several decades. In the case of reactions
postulated to occur via ionic intermediates in solution a great deal of mechanistic information may be obtained by a study of gas phase ion molecule processes for analogous systems. Ion cyclotron resonance spectroscopy has
been shown to be particularly useful in this respect since the techniques
of ICR single and double resonance and ion trapping are especially well
suited to the eludication of reaction pathways[1-3]. Since gas phase ion-
molecule reactions avoid the complicating of ion-solvent interactions the
intrinsic reactivity and thermochemical stability of reactants and products
may be determined. By further contrasting analogous solution and gas phase
processes it is frequently possible to assess the role of the solvent in
the reaction.

One system of interest is the solvolysis of acyl halides in acidic
media since in solution the reactions are sufficiently rapid - to preclu-
de use of many techniques of mechanistic study[4].

Results

The reactions of acyl halides, RCOX, with protonated species $R'MH^+$
can be devided into three general categories as illustrated by reactions

1 - 3. Reaction 1 is simple proton transfer and can be expected to

$$RCOX + R'MH^+ \longrightarrow RCOXH^+ + R'M \qquad (1)$$

$$RCOX + R'MH^+ \longrightarrow RCO^+ + R'M + HX \text{ (or } R'X + MH) \quad (2)$$

$$RCOX + R'MH^+ \longrightarrow RCOMR'^+ + HX \qquad (3)$$

occur whenever the proton affinity of the acyl halide is greater than that of the base R'M. Reaction 2 produces acylium ion and a pair of neutral products which may be either hydrogen halide, HX, plus base, R'M, or alkyl halide, R'X, plus a hydride MH. In some cases both pairs of neutral products are thermodynamically possible but in others only the latter pair may be exothermically produced. Reaction 3 is an addition-elimination reaction yielding a new protonated carboxylic acid derivative plus hydrogen halide which bears a strong resemblance in many cases to observed solvolysis reactions of acyl halides in acidic solution.

The possible reactions for each of the acyl halides with protonated substrates studied are summarized in Table 1. Included are the enthalpy change for each reaction, the total observed reaction rate constant for protonated substrate with acyl halide, and the experimentally determined product distribution.

1. Reactions of H_3O^+ and H_3S^+

From Table 1 it may be seen that the dominant reaction path for H_3O^+ and H_3S^+ with acyl halides is acetyl cation formation. Data for a mixture of H_2S and acetyl fluoride is shown in Fig. 1 and is typical for reactions between these species. The observed reactions occurring in this mixture confirmed by double resonance are summarized below. Similar reactions

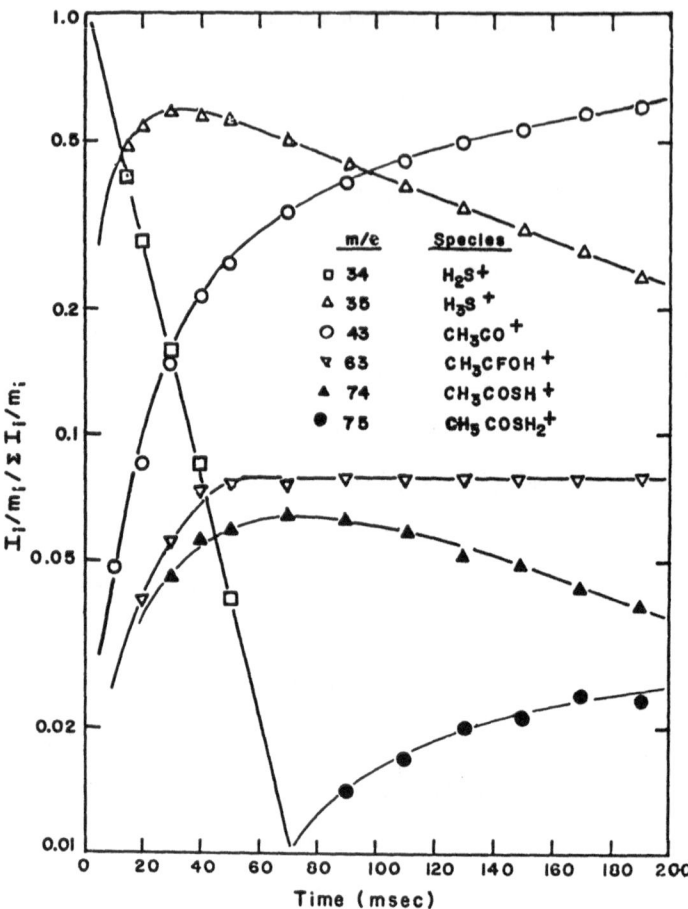

Fig. 1. Variation of relative ion abundances with reaction time in a 3:1 mixture of H_2S and acetyl fluoride at 15 eV and a total pressure of 10×10^{-6} torr.

are observed for H_3O^+ with acetyl fluoride and acetyl chloride and for H_3S^+ with acetyl chloride.

$$H_2S^+ + H_2S \longrightarrow H_3S^+ + SH \tag{4}$$

$$H_2S^+ + CH_3COF \longrightarrow CH_3COSH^+ + HF \tag{5}$$

$$H_3S^+ + CH_3COF \longrightarrow CH_3CO^+ + HF + H_2S \tag{6}$$

$$H_3S^+ + CH_3COF \longrightarrow CH_3COFH^+ + H_2S \tag{7}$$

$$CH_3COFH^+ + H_2S \longrightarrow (CH_3COSH)H^+ + HF \tag{8}$$

These results establish the proton affinities of both acetyl halides to be above that of H_2S, 172 kcal/mole.

In the reactions of H_3O^+ and H_3S^+ with benzoyl fluoride no protonated benzoyl fluoride is observed even though the proton affinity of benzoyl fluoride is established to be greater than that of H_2S. It may be seen from Table 1 that the decomposition of protonated benzoyl fluoride to benzoyl cation and HF is 3 kcal/mole exothermic. Thus it is likely that no protonated benzoyl fluoride is observed since exothermic proton transfer is immediately followed by unimolecular elimination of HF to produce the abundantly observed benzoyl cation, reaction 9.

$$H_3S^+ + C_6H_5COF \rightarrow H_2S + [C_6H_5CFOH^+]^* \rightarrow C_6H_5CO^+ + HF \tag{9}$$

2. Reactions of Protonated Alcohols

Protonated alcohols are found to react with acetyl fluoride and acetyl chloride to produce the corresponding protonated esters. The reactions observed in a mixture of acetyl fluoride and methanol are shown in Fig. 2 both as a function of increasing pressure and increasing reaction time to show the similarity of results from both types of experiment.

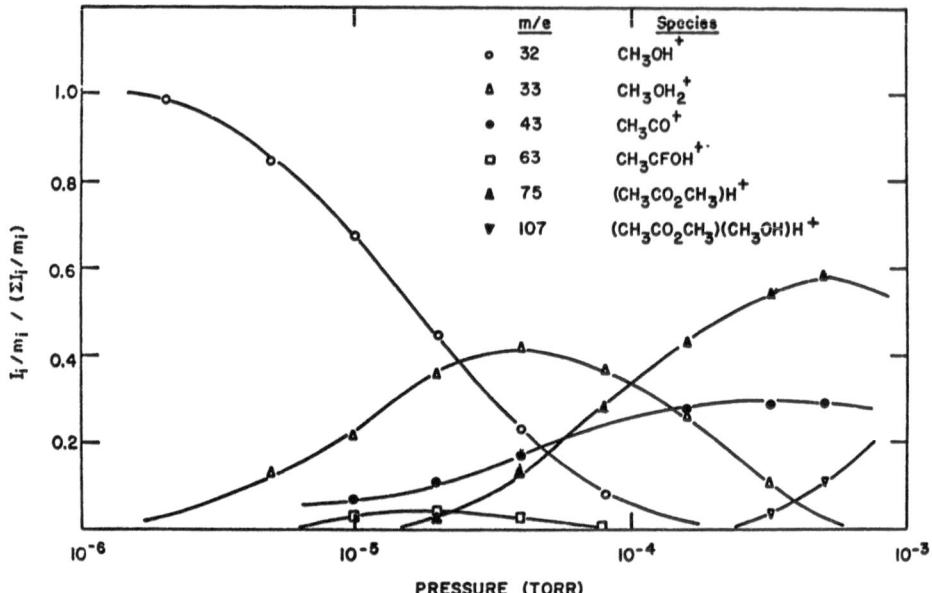

Fig. 2a. Variation of relative ion abundances with pressure in a 4.4:1 mixture of methanol and acetyl fluoride at 13 eV.

Reactions observed are summarized in equations 10 - 14.

$$CH_3OH^+ + CH_3OH \longrightarrow CH_3OH_2^+ + CH_3O \tag{10}$$

$$CH_2OH^+ + CH_3OH \longrightarrow CH_3OH_2^+ + CH_2O \tag{11}$$

$$CH_2OH^+ + CH_3COF \longrightarrow CH_3CO^+ + HF + CH_2O \tag{12}$$

$$CH_3OH_2^+ + CH_3COF \longrightarrow CH_3CO^+ + CH_3F + H_2O \tag{13}$$

$$CH_3OH_2^+ + CH_3COF \longrightarrow [CH_3CO_2CH_3]H^+ + HF \tag{14}$$

Reaction 13 is of particular interest since the neutral products have been inferred on the basis of thermochemical arguments and leads to important mechanistic considerations.

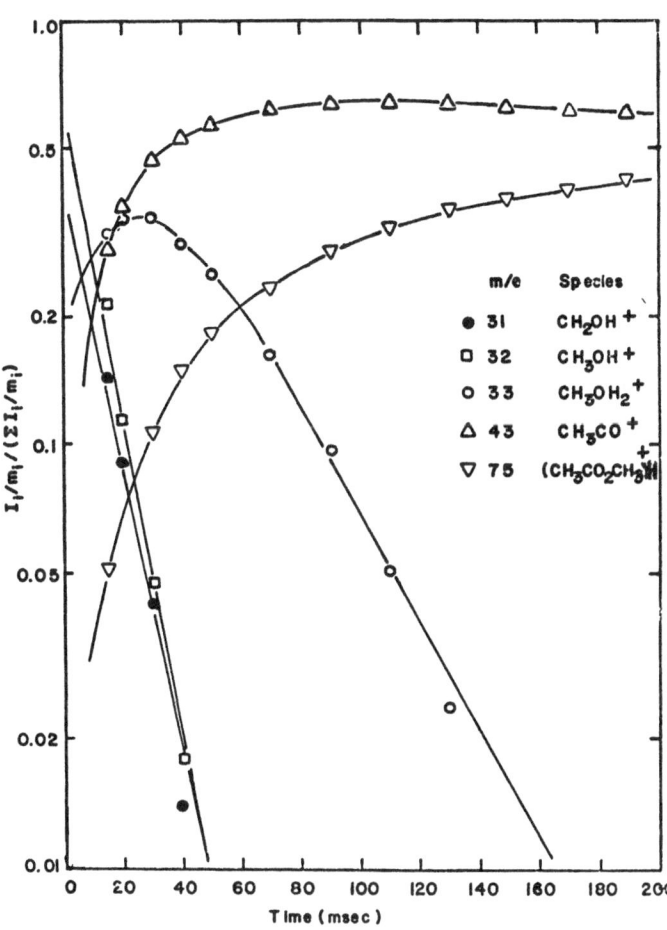

Fig. 2b. Variation of relative ion abundances with reaction time in a 2:1 mixture of methanol and acetyl fluoride at 14eV and a total pressure of 2.4x10^{-6} torr.

Similar results are observed in a mixture of ethanol with acetyl fluoride or acetyl chloride and methanol with acetyl chloride. In the case of ethanol and acetyl fluoride however another interesting product at m/e 87 is observed as a result of reaction of m/e 45, protonated acetaldehyde with the acetyl fluoride. This reaction is in principle similar to the acidic solvolysis reactions of protonated alcohols to yield protonated esters and the assumed product from protonated acetaldehyde is protonated biacetyl. This product does not appear in ethanol-acetyl chloride mixtures presumably because it does not compete favorably with acetyl cation formation. Typical data for ethanol-acetyl chloride mixtures is shown in Fig. 3.

In each of the cases where the presumed acidic solvolysis reaction is observed an alternative explanation of the reaction products is possible. In the gas phase reactions of Li^+ with acetyl bromide reaction 14' was observed which has the character of an acid induced elimination reaction producing an Li^+ bound to ketene[10].

$$Li^+ + CH_3COBr \longrightarrow Li...^+O=C=CH_2 + HBr \qquad (14')$$

In analogy to this reaction it was considered possible that an alternative explanation of reactions such as 14 might be 14a where the ionic product now is a proton bound dimer of methanol and ketene.

$$CH_3OH_2^+ + CH_3COF \longrightarrow CH_2=C=O...^+H(HOCH_3) + HF \qquad (14a)$$

Since the elimination of HF from acetyl fluoride is only about 27 kcal/mole endothermic if the binding energy of protonated methanol to acetyl fluoride were in excess of 27 kcal/mole an elimination of HF such as in

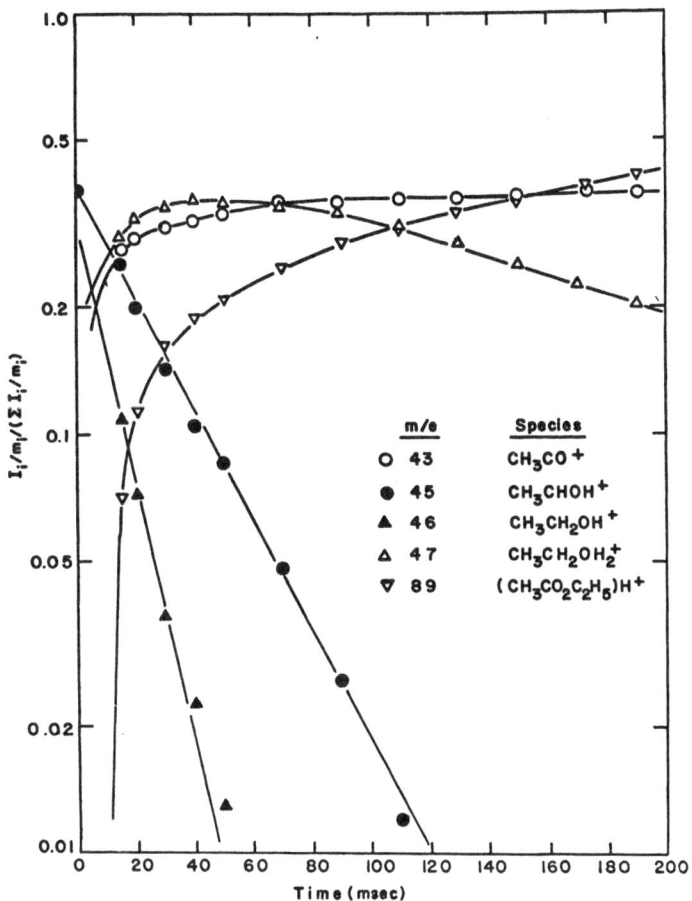

Fig. 3. Variation of relative ion abundances with reaction time in a 4:1 mixture of ethanol and acetyl chloride at 13 eV and a total pressure of 1.0×10^{-6} torr.

14a might be induced. In order to investigate this possibility the reactions occuring in a mixture of CD_3OD and CH_3COF were examined. In this case only reaction 14b was observed and not 14c.

$$CD_3OD_2^+ + CH_3COF \longrightarrow (CH_3CO_2CD_3)D^+ + DF \qquad (14b)$$

$$CD_3OD_2^+ + CH_3COF \overset{\times}{\longrightarrow} CH_2=C=O..H^+...(HOCH_3) + HF \quad (14c)$$

With benzoyl fluoride protonated methanol exhibited significantly different behaviour than the other protonated alcohols. Protonated methanol was found to react with benzoyl fluoride to produce only benzoyl cation while each the other protonated aliphatic alcohols also yielded protonated alkyl benzoates. Further experiments established that proton affinity of benzoyl fluoride was intermediate between methanol and ethanol and can be assigned a value of 184 ± 2 kcal/mole. Typical data for alcohol-benzoyl fluoride systems where acidic solvolysis occurs is shown in Fig. 4 for a mixture of ethanol and benzoyl fluoride. In addition to reactions similar to those previously observed a rather novel reaction product at m/e 106 was found to arise from reaction of the benzoyl fluoride molecular ion, m/e 124, with ethanol. A feasible explanation of this product is reaction 15.

$$C_6H_5COF^+ + CH_3CH_2OH \longrightarrow C_6H_5CHO^+ + CH_3F + CH_2O \text{ (or } HF + CH_3CHO)$$
$$(15)$$

3. Reactions of Protonated Ammonia and Amines

Ammonium ion and all alkyl ammonium ions were observed to be unreactive towards each of the acyl halides studied. The ion-molecule reactions occuring in a mixture of ammonia and acetyl chloride are shown in Fig. 5.

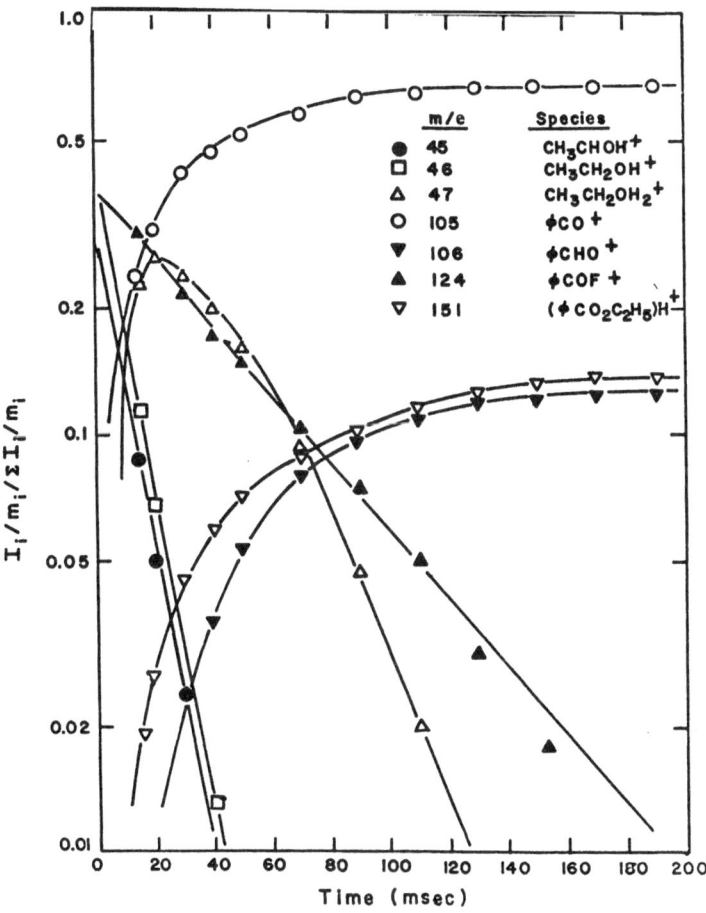

Fig. 4. Variation of relative ion abundances with reaction time in a 3.4:1 mixture of ethanol and benzoyl fluoride at 14 eV and a total pressure of 2.2×10^{-6} torr.

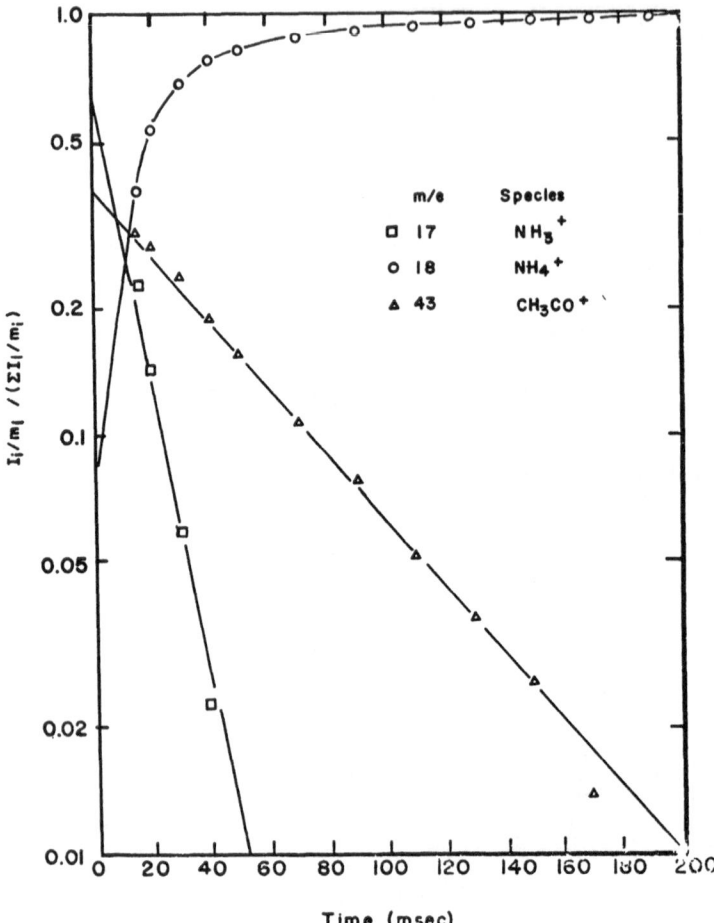

Fig. 5. Variation of relative ion abundances with reaction time in a 3.1:1 mixture of ammonia and acetyl chloride at 15 eV and a total pressure of 2.0x10^{-6} torr.

The only reaction of interest is proton transfer from the acyl cation to ammonia, reaction 16, in agreement with recent determination of the proton affinity of ketene which establish it to be less than ammonia.

$$CH_3CO^+ + NH_3 \longrightarrow NH_4^+ + CH_2CO \qquad (16)$$

Additional interesting results were obtained in mixtures of ammonia and amines with benzoyl fluoride. Since the ionization potential of benzoyl fluoride is comparable to ammonia it is necessary to operate under conditions where benzoyl fluoride molecular ion is also present. This ion initiates the sequence of reactions 17 and 18.

$$C_6H_5CFO^+ + NH_3 \longrightarrow C_6H_5CONH_2^+ + HF \qquad (17)$$

$$C_6H_5CONH_2^+ + NH_3 \longrightarrow C_6H_5CONH_3^+ + NH_2 \qquad (18)$$

$$NH_3^+ + C_6H_5COF \longrightarrow C_6H_5CFO^+ + NH_3 \qquad (19)$$

$$NH_3^+ + C_6H_5COF \longrightarrow C_6H_5CO^+ + HF + NH_2 \qquad (20)$$

In addition reactions 19 and 20 also occur. These reactions are illustrated in Fig. 6. Analogous behaviour is observed for methyl amine-benzoyl fluoride mixtures.

4. Reactions of Protonated Carboxylic Acids

The reactions of protonated formic acid with acetyl fluoride, Fig. 7, and acetyl chloride, Fig. 8, are noteworthy in that radically different products are formed in each case. With acetyl fluoride protonated formic acid reacts to produce a new ion at m/e 82 corresponding to protonated formic acetic anhydride, reaction 21. This reaction is analogous to the acidic solvolysis reactions observed above.

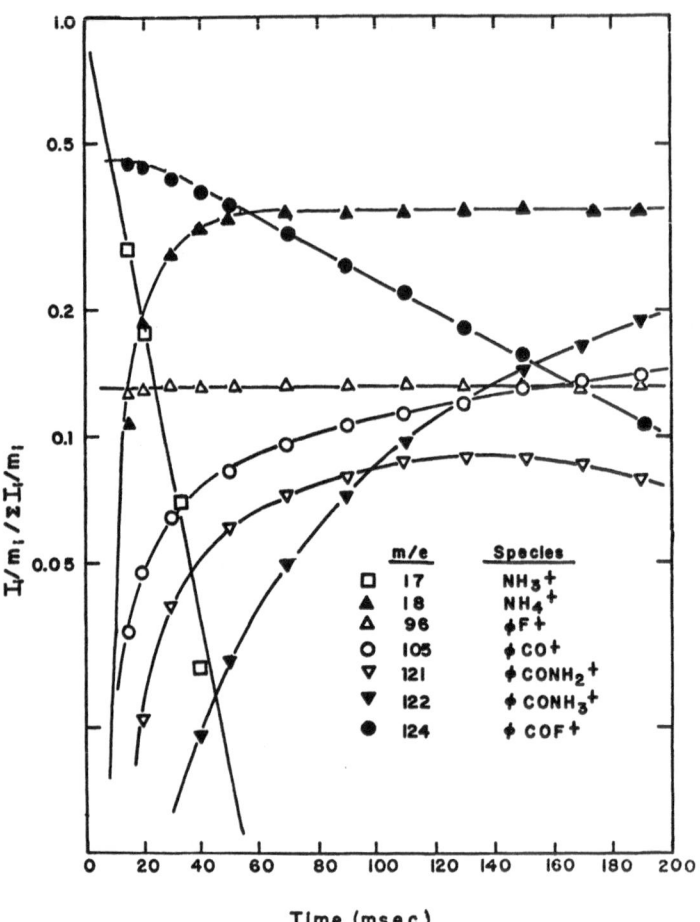

Fig. 6. Variation of relative ion abundances with reaction time in 4:1 mixture of ammonia and benzoyl fluoride at 14 eV and a total pressure of 2.0×10^{-6} torr.

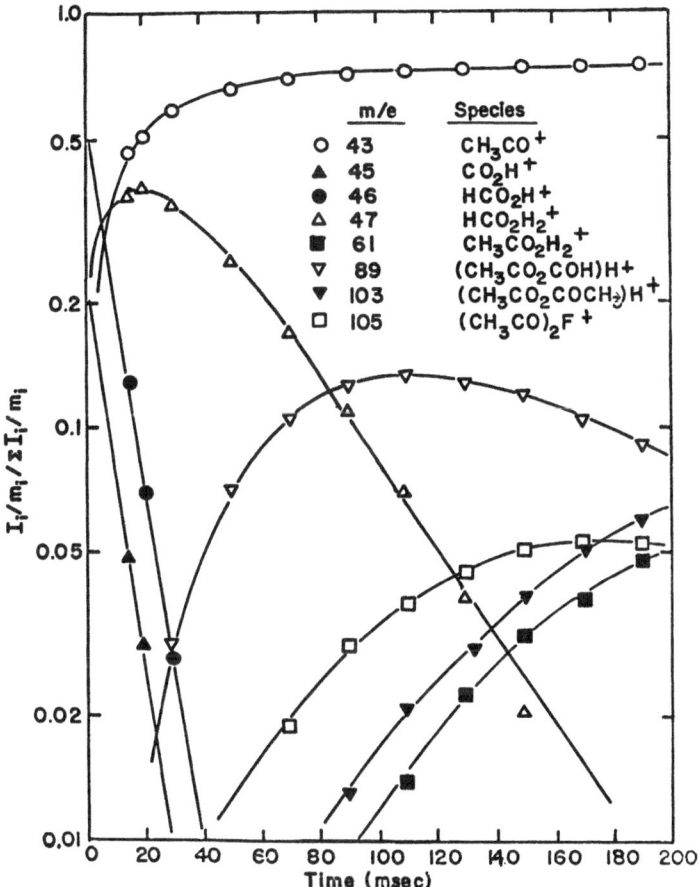

Fig. 7. Variation of relative ion abundances with reaction time in a 4:1 mixture of formic acid and acetyl fluoride at 17 eV and a total pressure of 2.0×10^{-6} torr.

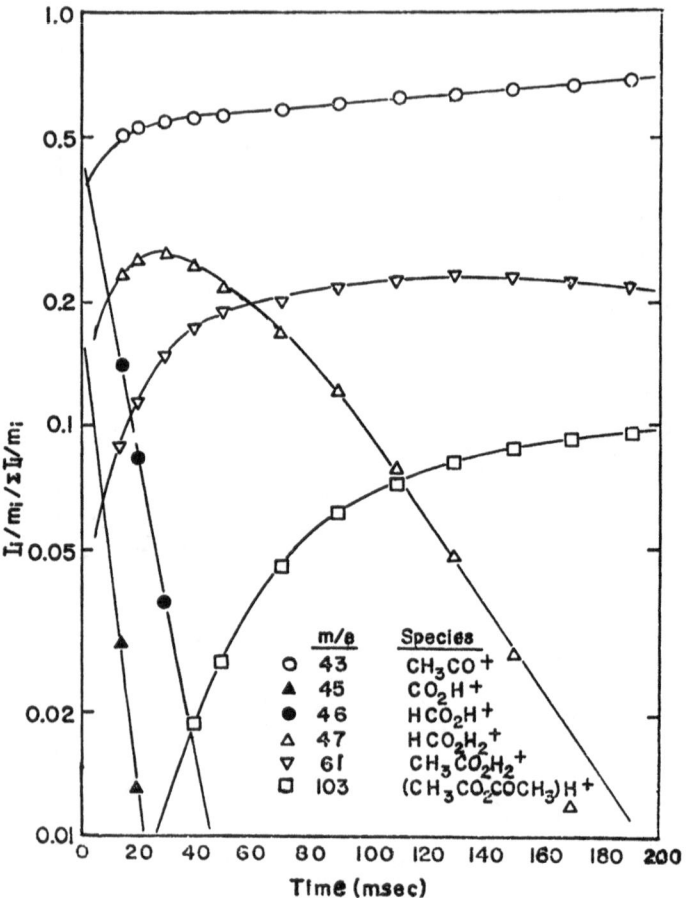

Fig. 8. Variation of relative ion abundances with reaction time in a 3:1 mixture of formic acid and acetyl chloride at 15 eV and a total pressure of 1.6×10^{-6} torr.

$$HCO_2H_2^+ + CH_3COF \longrightarrow (CH_3CO_2COH)H^+ + HF \qquad (21)$$

In addition products also appear at m/e 61 and 105 which may be explained by the sequence of reactions 22 - 24.

$$(HCO_2COCH_3)H^+ + CH_3COF \longrightarrow (CH_3CO)_2F^+ + HCO_2H \qquad (22)$$

$$(HCO_2COCH_3)H^+ + CH_3COF \longrightarrow ((CH_3CO)_2O)H^+ + HF + CO \qquad (23)$$

$$(CH_3CO)_2F^+ + HCO_2H \longrightarrow CH_3CO_2H_2^+ + CH_3COF + CO \qquad (24)$$

It is interesting to note that reactions 22 and 24 constitute a chain reaction leading to the overall decomposition of protonated formic acetic anhydride to protonated acetic acid plus carbon monoxide, reaction 25.

$$(HCO_2COCH_3)H^+ \longrightarrow CH_3CO_2H_2^+ + CO \qquad (25)$$

In the reactions of the protonated formic acid with acetyl chloride no protonated mixed anhydride is formed. Instead the immediate product of reaction is m/e 61, protonated acetic acid, reaction 26, followed by further reaction with acetyl chloride to produce the protonated symmetric anhydride, reaction 27.

$$HCO_2H_2^+ + CH_3COCl \longrightarrow CH_3CO_2H_2^+ + HCl + CO \qquad (26)$$

$$CH_3CO_2H_2^+ + CH_3COCl \longrightarrow ((CH_3CO)_2O)H^+ + HCl \qquad (27)$$

Thus although protonated acetic acid leads to anhydride formation as expected from the acetyl fluoride results protonated formic acid does not, contrary to expectation. Further experiments established the proton affinity of acetyl fluoride as intermediate between H_2S and formic acid and

of acetyl chloride as intermediate between formic acid and methanol.
These results have important mechanistic implications to be discussed below.

5. Reactions of Other Protonated Species

In order to test the generality of the addition-elimination reaction
1, a number of other protonated species were reacted with the acyl halides. CH_5^+, H_2Cl^+, and CH_3ClH^+ reacted only via proton transfer or acyl
cation formation. Protonated alkyl ethers and ketones were completely unreactive toward the acyl halides. In addition as seen from the data presented above protonated esters and anhydrides underwent no addition-elimination reactions with the acyl halides.

Discussion

From the above data it is evident that not all protonated species
will undergo the addition-elimination reaction with acyl halides. However careful examination reveals that when the following criteria are met
the acidic solvolysis reaction may be expected to occur.

1. The reaction must be exothermic. This is a general requirement
   of all ion-molecule reactions[6].

2. Proton transfer from the protonated "solvent" molecule to the
   acyl halide must be endothermic. This requirement is similar to
   that observed for gas phase nucleophilic displacement reactions[4].
   For example although reactions 28 and 29 are both exothermic, 28
   occurs while 29 does not since the competing reaction 30 is also
   exothermic.

$$C_2H_5ClH^+ \quad + \quad H_2O \quad \longrightarrow \quad C_2H_5OH_2^+ \quad + \quad HCl \qquad (28)$$

$$CH_3ClH^+ \quad + \quad H_2O \quad \longrightarrow \quad CH_3OH_2^+ \quad + \quad HCl \qquad (29)$$

$$CH_3ClH^+ \quad + \quad H_3O \quad \longrightarrow \quad H_3O^+ \quad + \quad CH_3Cl \qquad (30)$$

An exactly analogous situation may be seen for the acidic solvolysis reactions observed here. Though although both 31 and 32
are very exothermic only 31 is observed since 33 dominates in
reactions of protonated methanol with benzoyl fluoride.

$$CH_3CH_2OH_2^+ \quad + \quad C_6H_5COF \quad \longrightarrow \quad (C_6H_5CO_2C_2H_5)H^+ \quad + \quad HF \qquad (31)$$

$$CH_3OH_2^+ \quad + \quad C_6H_5COF \quad \longrightarrow \quad (C_6H_5CO_2CH_3)H^+ \quad + \quad HF \qquad (32)$$

$$CH_3OH_2^+ \quad + \quad C_6H_5COF \quad \longrightarrow \quad C_6H_5FOH^+ \quad + \quad CH_3OH \qquad (33)$$
$$\longrightarrow \quad C_6H_5CO^+ \quad + \quad HF \quad + \quad CH_3OH$$

In this situation where two exothermic reaction channels are
available from a common intermediate the channel involving the
more extensive molecular rearrangement is less favorable. For
similar reasons protonated formic acid undergoes addition-elimination with acetyl fluoride but not with acetyl chloride; H_3O^+
and H_3S^+ react only by proton transfer and ethanol and higher
alcohols all undergo the acidic solvolysis reaction.

3. The protonated solvent molecule must have at least two labile protons. This requirement arises from the observation that although
of protonated alcohols with acyl halides yield the corresponding
protonated esters, (for example reaction 14) protonated esters
are unreactive toward acyl halides. For example reaction 34 is
not observed even though it is 20 kcal/mole exothermic.

$$(CH_3)_2OH^+ \; + \; CH_3COF \; \longrightarrow \; (CH_3CO_2CH_3)H^+ \; + \; CH_3F \qquad (34)$$

Similarly no acidic solvolysis was ever observed from protonated species with less than two labile protons. It is important to note that the two labile protons need not to be bound to the same atom and need not to be equivalent. For example protonated formic acid reacts with acetyl fluoride to give protonated formic acetic anhydride, reaction 21, even though the requisite two protons are on different but equivalent oxygens. In addition protonated acetaldehyde is observed to react with acetyl chloride to produce protonated biacetyl, reaction 35, even though one of the labile protons is bound to oxygen and the other to carbon. Ab initio calculations indicate that the hydrogen bound has a formal positive charge of +0.412 while the hydrogen bound to the carbonyl

$$CH_3CHOH^+ \; + \; CH_3COF \; \longrightarrow \; ((CH_3CO)_2)H^+ \; + \; HF \qquad (35)$$

carbon has a charge of +0.289[7]. It is therefore reasonable to expect a certain degree of lability from this second hydrogen.

4. The protonated solvent must have at least one additional lone pair of electrons to participate in new bond formation. This requirement is indicated by the inability of protonated ammonia and amines to react with acyl halides to produce protonated amides even though the reactions are quite exothermic and in the case of the acetyl halides no other exothermic competing reaction channel of any kind exists. This criterion again has precedent if the ion-molecule reactions of alkyl halides, 8, 9, alcohols,

10 - 12, and amines, 13 - 15, are compared. While reactions 36

and 37 occur readily for protonated alcohols and alkyl halides

the analogous reaction, 38, of protonated amines is not observed

at all even though it is exothermic.

$$ROH_2^+ \ + \ ROH \quad \longrightarrow \quad R_2OH^+ \ + \ H_2O \qquad\qquad (36)$$

$$RXH^+ \ + \ RX \quad \longrightarrow \quad RXR^+ \ + \ HX \qquad\qquad (37)$$

$$RNH_3^+ \ + \ RNH_2 \quad \stackrel{}{\longrightarrow\!\!\!\!\!\!\times\!\!\!\!\!\longrightarrow} \quad R_2NH_2^+ \ + \ NH_3 \qquad\qquad (38)$$

This result is attributable to the lack of an available electron

pair on nitrogen in the proton bound dimer intermediate for the

reaction. In the alcohols and halides simultaneous bond breakage

and formation can occur in the intermediate since in these cases

the heteroatoms have one or two non bonding electron pairs avai-

lable to participate in formation of new bonds.

On the basis of the evidence presented above the probability is

high that if all four of these criteria are met an addition-elimination

reaction having the character of an acidic solvolysis will occur to an

observable extent. If other reaction channels are available the fraction

of protonated species reacting with acyl halide via this pathway may be

appreciably less than unity. In the cases examined above however wherever

the suggested criteria are met some acidic solvolysis is observed.

Consideration of the above criteria and appropriate thermochemical

data provides several clues on the details of the mechanims of several

of the reactions observed. On the basis of observation of reactions of

protonated molecules with n-donor bases the most likely initially formed

intermediate in the acidic solvolysis reaction is a linear proton bound

dimer[16,17]. This geometry is supported by several ab initio calculations
of model systems[18-20]. This intermediate is typified by structure I for
the methanol-acetyl fluoride system.

$$\text{CH}_3\text{C}\underset{F}{\overset{O.....H-O}{<}}\overset{H}{\underset{CH_3}{<}}+$$

In order to eliminate HF to produce protonated methyl acetate or to produce acetyl cation two further reaction intermediates are possible. The
first of these, structure II, is a tetrahedral structure, postulated by
solution organic chemists as the intermediate involved in the solvolysis
of acyl halides. Structure III is a bent proton bound dimer which is in general considered to be less stable than a linear proton bound dimer.
However Del Bene has recently carried out calculations of hydrogen bonded
dimers of bifunctional carbonyl compounds and has shown that in several
systems such as the dimer of urea and water two bent O-H...O and N-H...N
bonds may be formed[21]. This system is conceptually similar since a second
weaker hydrogen bond may form to the fluorine. Both II and III are adequate to explain the formation of protonated methyl acetate.

$$\text{H}_3\text{C-C}\underset{F}{\overset{OH}{<}}\overset{+}{\underset{H}{O}}\overset{CH_3}{<} \qquad \text{H}_3\text{C-C}\underset{F...H}{\overset{O...H}{<}}\overset{+}{>}\text{O-CH}_3$$

II   III

However an analysis of additional reactions which do and do not occur
provides a useful indication of which of II and III is more plausible.
In addition to protonated ester formation it is observed that protonated

methanol reacts with acetyl fluoride to produce acyl cation. The corresponding neutral products may be either methanol and HF or methyl fluoride and water. An analysis of thermochemical data reveals however that only the reaction associated with the latter pair of neutral products is exothermic. Elimination of CH_3F from II suggests that formation of protonated acetic acid would also be a feasible reaction since this reaction is exothermic by 22 kcal/mole. Since this product is in fact not observed structure III seems to be the more reasonable intermediate. Elimination of HF simultaneous with carbonyl carbon to alcohol oxygen bond formation facilely leads to the protonated ester. In addition rotation about the O-H...O bond allows for elimination of CH_3F and H_2O to yield acetyl cation.

The above conjecture is supported by mechanistic studies of esterification of acetic acid using deuterium and ^{18}O labelling carried out by Beauchamp[22]. In this case esterification occurs for all alcohols and acetic acids but by different mechanisms depending upon the relative proton affinities of the alcohol and the acid. The mechanistic conclusions drawn are outlined in Scheme 1. In this system the proton affinities of methanol and ethanol are less than acetic acid so that the mechanism involves proton transfer to the acid and formation of a tetrahedral intermediate followed by loss of H_2O. For isopropanol and tert-butanol the alcohol proton affinities are greater than acetic acid and the mechanism involves intramolecular R^+ transfer within the intermediate with loss of H_2O. The latter mechanism is exactly analogous to that proposed here for reaction of protonated species with acyl halides. The proton affinity of the acyl halide is less than the protonated species and a tetrahedral

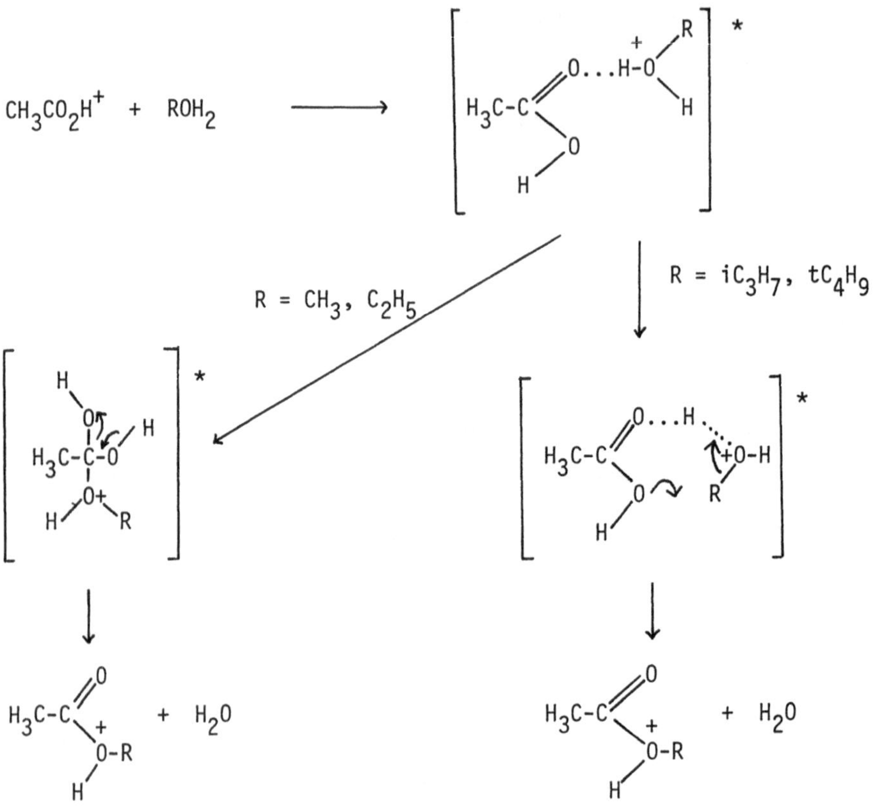

$$CH_3CO_2H^+ + ROH_2 \longrightarrow \left[H_3C-C \underset{O \diagdown H}{\overset{O \cdots H-O \diagup R}{\diagup\kern-1em\diagdown}} \right]^*$$

R = CH_3, C_2H_5

R = iC_3H_7, tC_4H_9

$$\left[\begin{array}{c} H \diagdown O \\ H_3C-C-O \\ H \diagup O^+ \diagdown R \end{array} \right]^*$$

$$\left[H_3C-C \underset{O \diagdown H}{\overset{O \cdots H}{\diagup\kern-1em\diagdown}} \; N^+O-H \; R \right]^*$$

$$H_3C-C \overset{O}{\underset{O-R}{\diagup\kern-1em\diagdown}} \; + \; H_2O \\ H$$

$$H_3C-C \overset{O}{\underset{O-R}{\diagup\kern-1em\diagdown}} \; + \; H_2O \\ H$$

Scheme 1

intermediate need not to be formed.Using the methanol-acetyl fluoride case as an example the mechanism can be illustrated by Scheme 2. The evidence and precedence all tend to support this as the most reasonable mechanism for acyl halide acidic solvolysis reactions.

It is also of interest to note that in certain cases where proton transfer from the protonated species to acyl halide is exothermic reac-

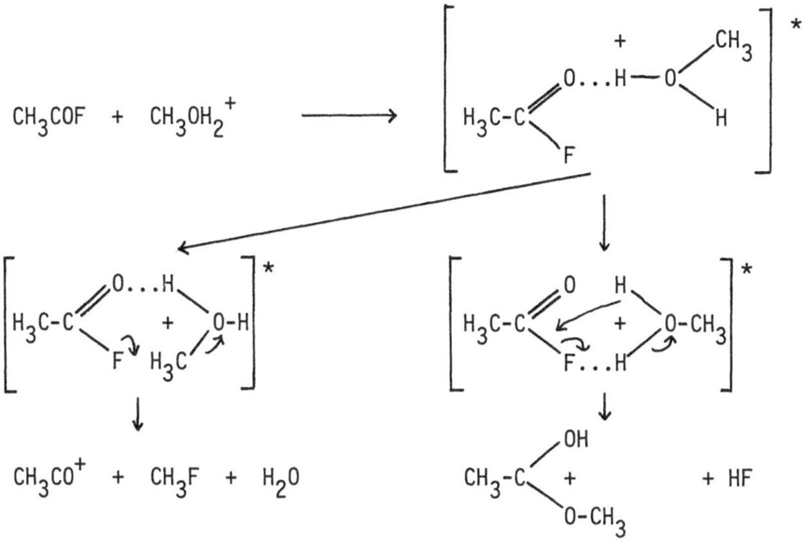

$$CH_3COF + CH_3OH_2^+ \longrightarrow \left[H_3C-C \overset{O\cdots H-O}{\underset{F}{\diagup}} \overset{+}{\diagdown} \overset{CH_3}{\underset{H}{\diagup}} \right]^*$$

$$\left[H_3C-C \overset{O\cdots H}{\underset{F}{\diagup}} \overset{+}{\underset{H_3C}{\diagup}} O-H \right]^* \longleftarrow$$

$$\downarrow$$

$$CH_3CO^+ + CH_3F + H_2O$$

$$\downarrow$$

$$\left[H_3C-C \overset{O \quad H}{\underset{F\cdots H}{\diagup}} \overset{+}{\diagdown} O-CH_3 \right]^*$$

$$\downarrow$$

$$CH_3-C \overset{OH}{\underset{O-CH_3}{\diagup}} + \quad + HF$$

Scheme 2

tions may occur which can be assumed to arise from a tetrahedral intermediate. For example the reactions of protonated formic acid with acetyl
fluoride and acetyl chloride are markedly different. In the case of acetyl fluoride the proton affinity is less than formic acid and the
acidic solvolysis product, protonated acetic-formic anhydride is formed.
This reaction is illustrated by Scheme 3. However the proton affinity
of acetyl chloride is greater than that of formic acid and the observed
reaction product, protonated acetic acid, must arise from a different
mechanism. Two resonable mechanisms presented by Schemes 4a and 4b are
possible.

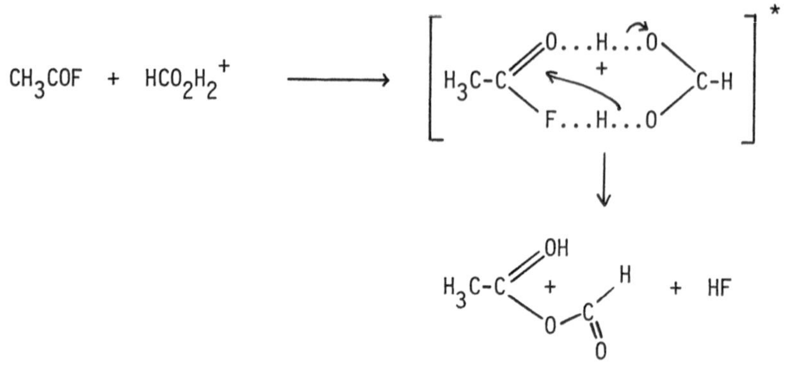

$$CH_3COF + HCO_2H_2^+ \longrightarrow$$

+ HF

Scheme 3

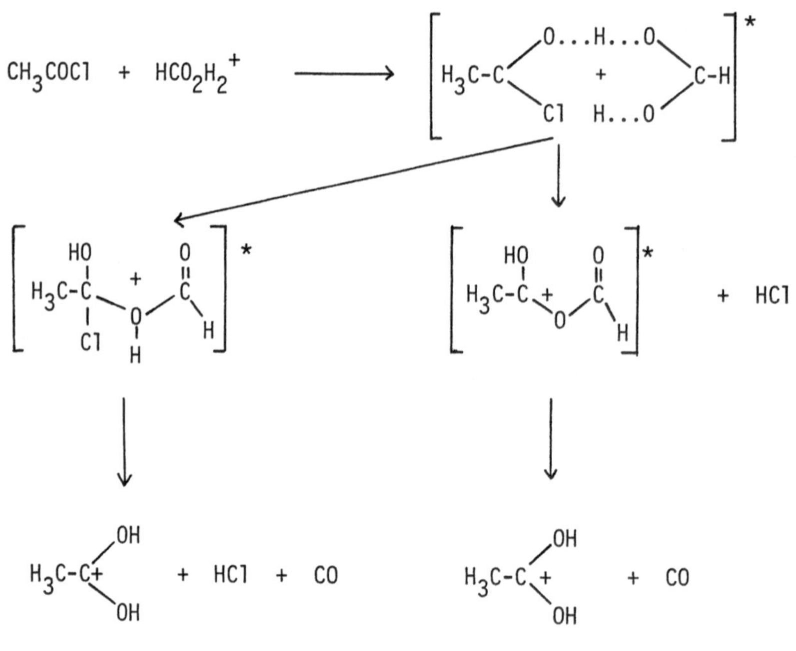

$$CH_3COCl + HCO_2H_2^+ \longrightarrow$$

+ HCl

+ HCl + CO

+ CO

Scheme 4a

Scheme 4b

Conclusion

The results and arguments presented above provide a rational basis for the general understanding of ion-molecule reactions of protonated species with acyl halides. The competition between various reaction channels may be understood given the criteria described for acidic solvolysis to occur and in cases of new reactants prediction of dominant reaction pathways should be possible. In addition the gas phase ion-molecule mechanism provides a first approximation for understanding the much more complex solvolysis reaction in solution.

Table 1. Kinetics and Thermochemistry of Acyl Halide Reactions

Reactants	Products	$\Delta H^{a,b}$ (kcal/mole)	Product Distribution	Total Rate Constant ($cm^3 molec^{-1} s^{-1}$)
$H_3O^+ + CH_3COF$	$CH_3CFOH^+ + H_2O$	-6	0.5	
	$CH_3CO^+ + HF + H_2O$	-5	0.5	$3x10^{-9}$
	$CH_3CO_2H_2^+ + HF$	-24	--	
$H_3O^+ + CH_3COCl$	$CH_3CClOH^+ + H_2O$	-11	--	
	$CH_3CO^+ + HCl + H_2O$	-8	1.0	$3.0x10^{-9}$
	$CH_3CO_2H_2^+ + HCl$	-27	--	
$H_3O^+ + C_6H_5COF$	$C_6H_5CFOH^+ + H_2O$	-16	0.5	
	$C_6H_5CO^+ + HF + H_2O$	-15	0.5	$3.2x10^{-9}$
	$C_6H_5CO_2H_2^+ + HF$	-31^b	--	
$H_3S^+ + CH_3COF$	$CH_3CFOH^+ + H_2S$	-3	0.2	
	$CH_3CO^+ + HF + H_2S$	-2	0.8	$8x10^{-10}$
	$CH_3C(OH)(SH)^+ + HF$	-23^b	--	
$H_3S^+ + CH_3COCl$	$CH_3CClOH^+ + H_2S$	-8	--	
	$CH_3CO^+ + HCl + H_2S$	-5	1.0	$1x10^{-9}$
	$CH_3C(OH)(SH)^+ + HCl$	-26^b	--	
$H_3S^+ + C_6H_5COF$	$C_6H_5CFOH^+ + H_2S$	-13	0.4	
	$C_6H_5CO^+ + HF + H_2S$	-12	0.6	$2x10^{-9}$
	$C_6H_5C(OH)(SH)^+ + HF$	-30^b	--	
$HCO_2H_2^+ + CH_3COF$	$CH_3CFOH^+ + HCO_2H$	+3	--	
	$CH_3CO^+ + HF + HCO_2H$	+4	--	$1.9x10^{-9}$
	$(CH_3CO_2COH)H^+ + HF$	-18^b	1.0	

Table 1 cont'd.

Reactants	Products	$\Delta H^{a,b}$ (kcal/mole)	Product Distribution	Total Rate Constant $(cm^3molec^{-1}s^{-1})$
$HCO_2H_2^+ + CH_3COCl$	$CH_3CClOH^+ + HCO_2H$	-2	--	
	$CH_3CO^+ + HCl + HCO_2H$	+1	--	$1.9 \times 10^{-9,c}$
	$(CH_3CO_2COH)H^+ + HCl$	-21^b	--	
$HCO_2H_2^+ + C_6H_5COF$	$C_6H_5CFOH^+ + HCO_2H$	-7	--	
	$C_6H_5CO^+ + HF + HCO_2H$	-6	--	
	$(C_6H_5CO_2COH)H^+ + HF$	-28^b	--	
$CH_3OH_2^+ + CH_3COF$	$CH_3CFOH^+ + CH_3OH$	+7	--	
	$CH_3CO^+ + CH_3F + H_2O$	-4	0.4	1.2×10^{-9}
	$(CH_3CO_2CH_3)H^+ + HF$	-25	0.6	
$CH_3OH_2^+ + CH_3COCl$	$CH_3CClOH^+ + CH_3OH$	+2	--	
	$CH_3CO^+ + CH_3Cl + H_2O$	-2	0.3	1.5×10^{-9}
	$(CH_3CO_2CH_3)H^+ + HCl$	-28	0.7	
$CH_3OH_2^+ + C_6H_5COF$	$C_6H_5CFOH^+ + CH_3OH$	-3	--	
	$C_6H_5CO^+ + HF + CH_3OH$	-2(-14)	1.0	2.0×10^{-9}
	$(C_6H_5CO_2CH_3)H^+ + HF$	-32	--	
$C_2H_5OH_2^+ + CH_3COF$	$CH_3CFOH^+ + C_2H_5OH$	+12	--	
	$CH_3CO^+ + HF + C_2H_5OH$	+13(+16)	--	1.5×10^{-9}
	$(CH_3CO_2C_2H_5)H^+ + HF$	-23	1.0	
$C_2H_5OH_2^+ + CH_3COCl$	$CH_3CClOH^+ + C_2H_5OH$	+7	--	
	$CH_3CO^+ + HCl + C_2H_5OH$	+10(+3)	--	8.3×10^{-10}
	$(CH_3CO_2C_2H_5)H^+ + HCl$	-26	1.0	

Table 1 cont'd.

Reactants	Products	$\Delta H^{a,b}$ (kcal/mole)	Product Distribution	Total Rate Constant ($cm^3 molec^{-1}s^{-1}$)
$C_2H_5OH_2^{+}+C_6H_5COF$	$C_6H_5CFOH^{+}+C_2H_5OH$	+2	--	
	$C_6H_5CO^{+}+C_2H_5OH+HF$	+3(+6)	--	2.3×10^{-9}
	$(C_6H_5CO_2C_2H_5)H^{+}+HF$	-29^b	1.0	
$CH_3CO_2H_2^{+}+CH_3COF$	$CH_3CFOH^{+}+CH_3CO_2H$	+12	--	
	$CH_3CO^{+}+HF+CH_3CO_2H$	+13	--	2×10^{-9}
	$(CH_3CO_2COCH_3)H^{+}+HF$	-17^b	1.0	
$CH_3CO_2H_2^{+}+CH_3COCl$	$CH_3CClOH^{+}+CH_3CO_2H$	+7	--	
	$CH_3CO^{+}+HCl+CH_3CO_2H$	+10	--	2×10^{-9}
	$(CH_3CO_2COCH_3)H^{+}+HCl$	-20^b	1.0	
$NH_4^{+}+CH_3COF$	$CH_3CFOH^{+}+NH_3$	+26	--	
	$CH_3CO^{+}+HF+NH_3$	+27	--	NR^a
	$(CH_3CONH_2)H^{+}+HF$	-7^b	--	
$NH_4^{+}+CH_3COCl$	$CH_3CClOH^{+}+NH_3$	+21	--	NR
	$CH_3CO^{+}+HCl+NH_3$	+24	--	
	$(CH_3CONH_2)H^{+}+HCl$	-4^b	--	
$NH_4^{+}+C_6H_5COF$	$C_6H_5CFOH^{+}+NH_3$	+16	--	NR
	$C_6H_5CO^{+}+HF+NH_3$	+17	--	
	$(C_6H_5CONH_2)H^{+}+HF$	-10^b	--	

a) Proton affinity data taken from Kebarle et al.[28], all other thermochemical data from ref.[29], or estimated on basis of group equivalent methods[20].

b) In cases where proton affinities have not been determined they are estimated on the basis of substituent effects in similar systems.

c) Values in parentheses indicate H for the alternative pair of neutral products.

References

1. D. Holtz, J. L. Beauchamp and S. D. Woodgate, J. Am. Chem. Soc. 92, 7484 (1970)

2. C. A. Lieder and J. I. Brauman, J. Am. Chem. Soc. 96, 4028 (1974)

3. S. Benezra, M. K. Hoffman and M. M. Bursey, J. Am. Chem. Soc. 92, 7501 (1970)

4. A. Kivinenin "The Chemistry of Acyl Halides", S. Patai, ed., Interscience, London, 1972

5. R. D. Wieting, R. H. Staley and J. L. Beauchamp, J. Am. Chem. Soc. 97, 924 (1975)

6. J. L. Franklin, J. Chem. Educ. 40, 284 (1963)

7. A. C. Hopkinson and I. G. Csizmadia, Can. J. Chem. 52, 546 (1974)

8. J. L. Beauchamp, D. Holtz, S. D. Woodgate and S. L. Patt, J. Am. Chem. Soc. 94, 2798 (1972)

9. R. J. Blint, T. B. McMahon and J. L. Beauchamp, J. Am. Chem. Soc. 96, 1269 (1974)

10. M. S. B. Munson, J. Am. Chem. Soc. 87, 5313 (1965)

11. L. W. Sieck, F. P. Abrahamson and J. H. Futrell, J. Chem. Phys. 45, 2859 (1966)

12. J. M. S. Henis, J. Am. Chem. Soc. 90, 844 (1968)

13. M. S. B. Munson, J. Phys. Chem. 70, 2034 (1966)

14. L. Hellner and L. W. Sieck, Int. J. Chem. Kinet. 5, 177 (1973)

15. J. M. Brupbacher, C. J. Eagle and E. Tschaikow Roux, J. Phys. Chem. 79, 671 (1975)

16. A. J. Cunningham, J. D. Payzant and P. Kebarle, J. Am. Chem. Soc. 94, 7627 (1972)

17. P. Kebarle in "Ion-Molecule Reactions", J. L. Franklin ed., Plenum
    Press, New York, 1972

18. P. A. Kollman and L. C. Allen, J. Am. Chem. Soc. 92, 6101 (1970)

19. M. D. Newton and S. Ehrenson, J. Am. Chem. Soc. 93, 4971 (1971)

20. P. Lerlet, S. D. Peyerimhoff and R. J. Buenker, J. Am. Chem. Soc.
    94, 8301 (1972)

21. J. E. DelBene, J. Chem. Phys. 63, 4666 (1975)

22. J. L. Beauchamp in "Interactions Between Ions and Molecules", P. Aus-
    loos ed., Plenum Press, New York, 1975

23. R. Yamdagni and P. Kebarle, J. Am. Chem. Soc. 98, 1320 (1976)

24. J. L. Franklin, J. G. Dillard, H. M. Rosenstock, T. J. Herron and
    K. Draxl"Ionization Potentials, Appearance Potentials and Heats of
    Formation of Gaseous Positive Ions", NSRDS-NBS 26, U.S.Gov't Printing
    Office, Washington D. C., 1969

25. S. W. Benson and H. E. O'Neal, "Kinetic Data on Gas Phase Unimolecu-
    lar Reactions", NSRDS-NBS 21, U.S. Gov't Printing Office, Washington
    D. C., 1970

Positive and Negative Ionic Reactions at the Carbonyl
Bond in the Gas Phase

Fritz S. Klein and Zeev Karpas
Isotope Department
Weizmann Institute of Science
Rehovot, Israel

Introduction

Ion molecule interactions have been studied from many different aspects, such as the reaction kinetics of single systems, types and mechanism of reactions, the chemistry of specific ions with various groups of compounds etc.[*]. Most of these studies concentrated on either the positive or the negative ion chemistry.

The chemical behaviour of a specific bond, isolated as far as possible from neighboring effects, with respect to positive and negative ions has up to now, as far as we know, not been investigated specifically. It may prove useful to have such kind of a study in connection with investigations of various types of plasma, electrical discharge phenomena and of some kind of gas lasers.

A beginning in this direction of investigation we wish to make with the present work which constitutes a review of studies by our group, published over the last four years. The purpose was to study the ion chemistry of the carbonyl bond in compounds of the type X_2CO, where X is H, F or Cl, with ions generated by electron impact on the parent molecule. Effects on the

[*]For general reference: Interaction Between Ions and Molecules,

Edit. P. Ausloos, Plenum Press, New York and London, 1974

reactivity and the reaction mechanism were studied.

Positive ion - molecule chemistry

Formaldehyde is the most prolific in its ion molecule interactions [1].
On electron impact in the energy range of 30 to 50 eV formaldehyde forms
the parent ion CH_2O^+ and the fragment ions CHO^+, CO^+, CH^+ and C^+. All
these ions with the exception of CO^+ react with the neutral CH_2O. At low
neutral gas densities only the two fastest reactions were observed:

$$CH_2O^+ + CH_2O \rightarrow CH_3O^+ + CHO \qquad (1)$$

and
$$CHO^+ + CH_2O \rightarrow CH_3O^+ + CO. \qquad (2)$$

These reactions are proton exchanges. The structure of the product ion is
therefore $H_2C = OH^+$. This has been shown by the double resonance test
(see Appendix) on isotopically substituted formaldehyde: By observing in
a 1:1 mixture of $CH_2^{18}O$ and $CH_2^{16}O$ the product ion, $CH_3^{18}O^+$, and irradiating all isotopic reactant ions of the type CH_2O^+ and CHO^+, only the ions
$CH_2^{16}O^+$ and $CH^{16}O^+$ gave positive double resonance signals in $CH_3^{18}O^+$,
showing that a proton is transferred from the reactant ion $CH_2^{16}O^+$ or
$CH^{16}O^+$ to the neutral $CH_2^{18}O$. Similar by observing the product ion CHD_2O^+
in a 1:1 reaction mixture of deuterated formaldehyde, CD_2O, and formaldehyde, CH_2O, and sweeping the irradiating oscillator frequency over all
reacting ions, only the following proton transfers were observed:

$$CH_2O^+ + CD_2O \rightarrow CD_2OH^+ + CHO$$

and
$$CHO^+ + CD_2O \rightarrow CD_2OH^+ + CO.$$

Two secondary ions, CH_3O^+ and CH_2DO^+ gave also a positive double
resonance signal. This indicates that isotope exchange occurs according

to:

$$CH_3O^+ \;+\; CD_2O \;\longrightarrow\; CD_2OH^+ \;+\; CH_2O$$

and $\quad CH_2DO^+ \;+\; CD_2O \;\longrightarrow\; CD_2OH^+ \;+\; CH_2O$

which clearly also proceeds by proton transfer.

The rate constant ratio k_1/k_2 of reactions 1 and 2 has been determined by the relative ionization efficiency method, Fig. 1:

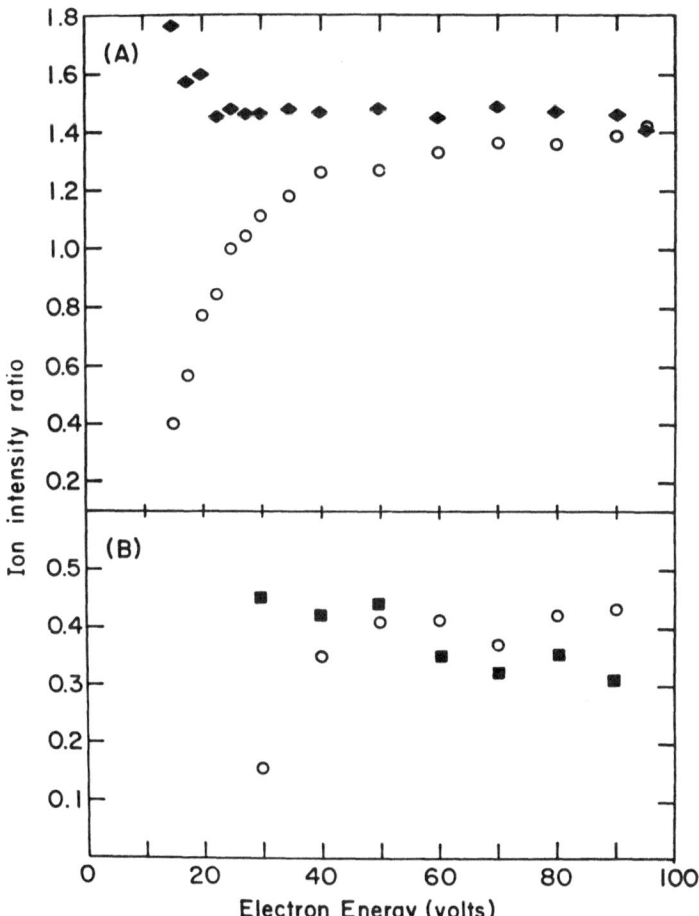

Fig. 1. Ion intensity ratios. A. CHO^+/CH_2O^+ (○) and $10 \times CH_3O^+/$ $(CH_2O^+ + CHO^+)$ (◆), B. CH^+/CH_2^+ (○) and $CH_3^+/(CH_2^+ + CH^+)$ (■), as a function of electron energy.

$$k_1/k_2 \; = \; 2.7 \pm 0.8 \text{ at room temperature.}$$

A second type of ion-molecule interactions, namely atom abstraction reactions, have been observed at intermediate reactant gas densities (3×10^{10} to 5×10^{11} molec. cm^{-3}). These reactions, which are about one order of magnitude slower than the proton transfer reactions, are:

$$C^+ + CH_2O \rightarrow CH_2^+ + CO, \tag{3}$$

$$CH_2^+ + CH_2O \rightarrow CH_3^+ + CHO, \tag{4}$$

and

$$CH^+ + CH_2O \rightarrow CH_3^+ + CO. \tag{5}$$

The double resonance test with a $CH_2O - CD_2O$ mixture again verified the competition of reactants in reactions (4) and (5). The rate constant ratio found was:

$$k_4/k_5 \; = \; 2.0 \pm 1.2 \; \text{ at room temperature.}$$

The product ion CH_4O^+ is formed predominantly by CH_2O^+:

$$CH_2O^+ + CH_2O \rightarrow CH_4O^+ + CO. \tag{6}$$

Double resonance could not detect any proton transfer contribution from CH_3O^+.

The tertiary reaction

$$CH_3O^+ + CH_2O \rightarrow CH_5O^+ + CO \tag{7}$$

is relatively slow (see Table 1) and probably because of the necessary isomerization from a protonated formaldehyde to a methoxy ion in order to give a protonated methanol, $CH_3OH_2^+$.

At the highest reactant gas densities studied (above 5×10^{11} molec. cm^{-3}) a third type of reaction mechanisms - complex formation - was observed. The secondary ion, protonated formaldehyde, CH_2OH^+, which at these pressures is the most abundant ion, reacts with CH_2O to give:

$$CH_3O^+ + CH_2O \rightarrow C_2H_5O_2^+. \tag{8}$$

This reaction channel has been verified by the double resonance test using a $CH_2{}^{18}O$ - $CH_2{}^{16}O$ mixture. All three isotopic products $C_2H_5{}^{16}O_2$, $C_2H_5{}^{16}O^{18}O$ and $C_2H_5{}^{18}O_2$ were observed in the expected statistical ratio. Fragment ions of this complex containing two carbon atoms were formed according to:

$$C_2H_5O_2^+ \rightarrow C_2H_5O^+ + O \tag{9}$$
$$\rightarrow C_2H_4O^+ + HO \tag{10}$$
$$\rightarrow C_2H_3O^+ + H_2O \tag{11}$$
$$\rightarrow C_2H_2O^+ + H + H_2O \tag{12}$$
$$\rightarrow C_2HO^+ + H_2 + H_2O \tag{13}$$
$$\rightarrow C_2O^+ + H_2 + H_2O + H. \tag{14}$$

The raction rate constants of the main reactions were determined by the graphic method of Anicich and Bowers [2]. Absolute values of these constants were established by comparing the rate constant of reaction 1 with that of the protonation of methane [3], (Fig. 2). The rate constants of the ion molecule reactions of formaldehyde are summarized in Table 1. The experimental rate constant of reaction 1 can be compared with values predicted by the various theories [4]:

A. Induced dipole   : $k_I = 0.77 \times 10^{-9}$ cm^3 $molecule^{-1}sec^{-1}$

B. Locked in dipole   : $k_L = 5.59$

C. ADO   : $k_A = 1.98$

D. Experimental value : $k_X = 1.64$

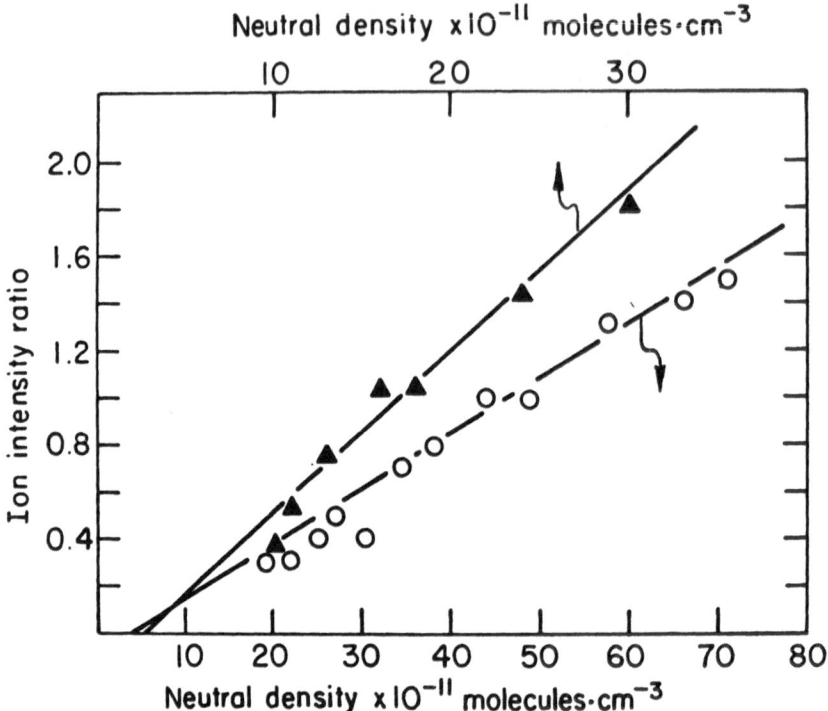

Fig. 2. Ion intensity ratios $CH_5(CH_5^+ + CH_4^+)$ (▲) and $CH_3O^+/(CH_3O^+$ $+ CH_2O^+ + CHO^+)$ (○), as a function of neutral molecule density of methane and formaldehyde, respectively.

Considering the experimental error the agreement with the "average dipole orientation" model is good.

A general reaction scheme for the formaldehyde system is given in Figure 3.

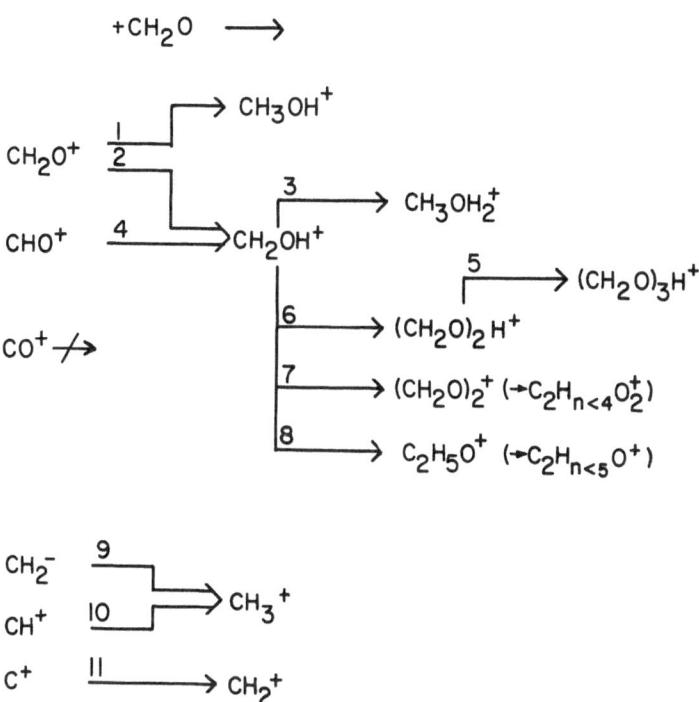

Fig. 3. Ion-molecule reaction scheme in the formaldehyde system.

Formyl fluoride, HFCO, reacts in many respects similar to formal-dehyde, still the fluorine atom increases the electron affinity of the molecule and affects its reactivity [5]. On electron impact formyl fluoride is ionized forming a parent ion, $HFCO^+$, and the fragment ions HCO^+, FCO^+, CO^+, CFH^+ and CF^+. The ions C^+, CH^+, O^+ and F^+ are formed with intensities of less than 1% of the total ion intensity.

The two dominant reactions are again proton transfers:

$$HFCO^+ + HFCO \rightarrow HFCOH^+ + FCO \qquad (15)$$

and $\qquad$ CHO^+ + HFCO $\rightarrow$ $HFCOH^+$ + CO. $\qquad$ (16)

The rate constant ratio of these reactions found was:

k_{15}/k_{16} = 3.6 $\pm$ 0.8.

The atom abstraction reactions observed in this system were:

$$CF^+ + HFCO \rightarrow CFH^+ + FCO \qquad (17)$$

and the formation of a tertiary protonated fluoromethanol:

$$HFCOH^+ + HFCO \rightarrow CHF_2OH_2^+ + CO \qquad (18)$$

in analogy to $CH_3OH_2^+$ of formaldehyde.

At high reactant densities complex formation is again observed:

$$HFCOH^+ + HFCO \rightarrow (HFCO)_2H^+ \qquad (19)$$

This dimer ion decomposes to give the fragment ions:

$$(HFCO)_2H^+ \rightarrow C_2H_2F_2O_2^+ + H \qquad (20)$$
$$\rightarrow C_2H_3FO_2^+ + F \qquad (21)$$
$$\rightarrow C_2H_3F_2O^+ + O \qquad (22)$$

The reaction rate constants of the HFCO system (reactions 15 to 22) are summarized in Table 2.

The carbonyl halides: F_2CO, FClCO and Cl_2CO [6] gave on electron impact positive ions of the type XCO^+, CX_2^+, CX^+, X^+ and CO^+, but even at the highest reactant gas densities ($\sim 10^6$ molec. cm^{-3}) no positive ion molecule reactions were observed. This puts an upper limit of 10^{-12} cm^3 molec^{-1} sec^{-1} for the rate constant of any reaction of this type.

Negative ion-molecule chemistry

The carbonyl halides: carbonyl fluoride, F_2CO, carbonyl chlorofluoride, FClCO, and phosgene, Cl_2CO exhibit a quite diverse negative ion-molecule chemistry [6]. On impact with electrons at low energies (0 to 10 eV) all three compounds form negative ions of the type XCO^-, X_2^- and X^- (for X = F or Cl) by dissociative electron attachment, see Table 3.

In phosgene only $ClCO^-$ and Cl_2^- have been observed to react with the neutral parent molecule:

$$ClCO^- + Cl_2CO \rightarrow Cl_3CO^- + CO \qquad (23)$$

and $\qquad Cl_2^- + Cl_2CO \rightarrow Cl_3CO^- + Cl \qquad (24)$

Both reactions proceed by halide ion transfer. This could be demonstrated in the case of reaction 24 by the double resonance test applied to the natural chlorine isotope abundance. When observing the product ion $^{35}Cl_2^{37}ClCO^-$, the intensity ratio of the double resonance signals for the reactant ions $^{35}Cl_2^-$, $^{35}Cl\ ^{37}Cl^-$ and $^{37}Cl_2^-$ found was: 6:5:1, which is exactly the ratio predicted for chloride ion transfer. The production of Cl_3CO^- by complex formation should have given an intensity ratio of 10:13:1 (see Appendix).

A second type of reaction forming secondary Cl^- was the collision induced decomposition of:

$$ClCO^- + Cl_2CO \rightarrow Cl^- + CO + Cl_2CO \qquad (25)$$

and $\qquad Cl_2^- + Cl_2CO \rightarrow Cl^- + Cl + Cl_2CO. \qquad (26)$

Carbonyl fluoride, F_2CO, similar to phosgene, forms the ions FCO^-, F_2^- and F^-. FCO^- and F_2^- react by halide ion transfer:

$$FCO^- + F_2CO \rightarrow F_3CO^- + CO \qquad (27)$$

and $\qquad F_2^- + F_2CO \rightarrow F_3CO^- + F. \qquad (28)$

Since the X_3CO^- ion in this case is far more stable than the analoge ion in phosgene no collision induced X^- ions were observed.

At high reactant density a reaction involving the secondary ion F_3CO^- was observed:

$$F_3CO^- + F_2CO \rightarrow F_5CO^- + CO \qquad (29)$$

This reaction is assumed to proceed by atom abstraction giving the ionic structure $(F_3CO \cdot F_2)^-$.

In carbonyl chlorofluoride the two product ions F_2ClCO^- and FCl_2CO^- were observed:

$$FCO^- + FClCO \rightarrow F_2ClCO^- + CO; \qquad (30)$$
$$ClCO^- + FClCO \rightarrow FCl_2CO^- + CO; \qquad (31)$$
$$ClF^- + FClCO \rightarrow F_2ClCO^- + Cl; \qquad (32)$$
and $$ClF^- + FClCO \rightarrow FCl_2CO^- + F. \qquad (33)$$

A double resonance test performed on reactions 32 and 33, which have identical reactants, gave only a weak signal for reaction 32 because this reaction is about 10 kcal.mole^{-1} less exothermic than the second. Table 4 summarizes the reaction rate constants of the carbonyl halides.

Studying binary mixtures of the three carbonyl halides it was possible to determine the relative chloride ion transfer rates for the three compounds by measuring the relative rate ratios : $k_{F_2CO}/k_{ClFCO} = 0.33 \pm 0.05$; $k_{F_2CO}/k_{Cl_2CO} = 0.73 \pm 0.05$. The order of increasing transfer rates is therefore: $F_2CO < Cl_2CO < ClFCO$. The formation of the product ions X_3CO^- in these binary mixtures by the reactants XCO^- and X_2^- permitted also the estimation of the heat of formation of these product ions. The results are summarized in Table 5.

Formyl fluoride [5]. At high reactant density only the slow reaction:

$$FCO^- + HFCO \rightarrow HF_2CO^- + CO \qquad (34)$$

was observed, its rate constant $k_{34} = 2 \times 10^{-12}$ cm^3 molec^{-1}s^{-1}. Using a mixture of HFCO and F_2CO and producing F^- by dissociative electron attachment at

F_2CO, this ion reacted according to:

$$F^- + HFCO \rightarrow FCO^- + HF$$

This reaction channel was verified by the double resonance test.

The fact that F^- is not observed in the mass-spectrum of HFCO can now be explained by the strong scavenging effect on any F^- ions formed initially.

Formaldehyde[7] does not form negative ions by electron impact, but HCO^- can be produced by proton abstraction:

$$NH_2^- + H_2CO \rightarrow HCO^- + NH_3, \tag{35}$$

and similarly to readtion (34):

$$HCO^- + H_2CO \rightarrow H_3CO^- + CO. \tag{36}$$

Positive as well as negative ion transfer reactions occur in mixtures of formaldehyde and carbonyl halides[5]. Thus in a mixture of H_2CO and F_2CO the main reactions are: Proton transfer

$$H_2CO^+ + F_2CO \rightarrow F_2COH^+ + HCO \tag{37}$$

and $\quad\quad H CO^+ + F_2CO \rightarrow F_2COH^+ + CO \tag{38}$

and fluoride ion transfer:

$$FCO^- + H_2CO \rightarrow H_2FCO^- + CO \tag{39}$$

$$F_2^- + H_2CO \rightarrow H_2FCO^- + F \tag{40}$$

In a mixture of formaldehyde and phosgene the proton transfers are:

$$H_2CO^+ + Cl_2CO \rightarrow Cl_2COH^+ + HCO \tag{41}$$

and $\quad\quad HCO^+ + Cl_2CO \rightarrow Cl_2COH^+ + CO, \tag{42}$

whereas the chlorine ion transfers are:

$$ClCO^- + H_2CO \rightarrow H_2ClCO^- + CO \tag{43}$$

$$Cl_2^- + H_2CO \rightarrow H_2ClCO^- + Cl. \tag{44}$$

Conclusions

The ion chemistry of the carbonyl bond of the type X_2CO has been found to be quite complex. The following general features can be listed:

1. The fastest reactions occur when the ionic reactant contains a proton: proton transfer, or when

2. the ionic reactant contains a halogen atom: halide ion transfer.

3. Hydrogen and/or halogen attached to the carbon of the neutral reactant are abstracted by the ionic reactant.

4. Complex formation occurs predominantly with protonated parent ions and any of the neutral reactants.

It has been shown in a related investigation [10] that, especially in halide ion transfer reactions, the electronegativity and the volume of the reactant group has a strong effect on the reactivity of a given reactant. In a similar fashion the channel of formation and the configuration of the complex can be affected by these properties. A four centered intermediate has been proposed to account for the relative rates of complex formation of mixtures of various carbonyl compounds of the type X_2CO ↑↑ and the fragmentation pattern of the complex (reactions 9 to 14 and reactions 20 to 22).

In summary the carbon of the carbonyl bond is found to be the site of electrophilic attach whereas the nucleophilic reaction center is on the oxygen atom.

Appendix: Experimental

Instrument: The ion cyclotron mass spectrometer used in these studies was a Varian V-5900 Syrotron equipped with a standard "flat" cell. A grid pla- ced between filament and electron beam entrance to the source region per- mitted operation in the pulsed beam mode. The pressure of the reacting gas was measured by the ion current in the Vac-Ion pump of the ICR mass spectrometer. This pump current was calibrated with a Veeco-Ionization gauge while the magnet field was turned off (to prevent damage to the gauge head filament by vibration).

Materials:

1. Formaldehyde, H_2CO, was obtained from paraformaldehyde (purum grade, Fluka) by vacuum distillation at $115^\circ C$ and subsequent trap to trap distillation of the product at $-80^\circ C$.

2. Deutero-formaldehyde, D_2CO, was produced by the same metho as formal- dehyde, starting from perdeutero formaldehyde (99 atom % D, Merck).

3. ^{18}O-formaldehyde, $H_2C^{18}O$, was obtained the same way as H_2CO. ^{18}O-para- formaldehyde was obtained by equilibrating normal paraformaldehyde with an excess ^{18}O-water in a closed ampoule at $150^\circ C$ for 90 min.

4. Formyl fluoride, HFCO, was prepared by the procedure of Olah and Kuhn[8] from   benzoyl chloride added to a mixture of anhydrous formic acid and potassium hydrogen fluoride. Distilling off the formyl fluoride, the product was purified by trap to trap distillation.

5-6. Carbonyl fluoride, F_2CO, and Carbonyl chlorofluoride, FCICO, (PCR Inc., Gainsville, Fla.) were purified by repeated freeze-pump-thaw cycles. Carbonyl fluoride was stored in stainless steel containers.

7. Phosgene, Cl_2CO (Matheson Co.) was purified similarly to carbonyl
    fluoride.

The double resonance test for determining the reaction mechanism is performed as follows [6] : A suitable isotopic product ion is chosen for
mon toring so that its mass number is unique (no overlap by other species
present). All isotopic species of the ionic reactant are scanned by the
irradiating oscillator and the relative double resonance signals are determined. The relative signal intensities are then a simple function of
the isotope abundances of the reactants and the statistical probability
of the product formation by a given mechanism (neglecting possible isotope effects).

Rate constant ratio determination of competing reactions (reactants)[1].
By   varying   the energy of the impact electrons over a limited range
the relative abundance of competing reactant ions can be changed appreciably. Observation of the product ion intensity as a function of the
reactant ion ratio permits the evaluation of the required rate constant
ratio by simple arithmetic. This method proved to be more accurate and
reliable than the two methods employed in earlier studies [9];

A. The reactant ejection method which often affects not only the reactant
to be ejected but also the second reactant and

B. The limiting electron energy method, which operates at two extreme
energies, a low energy where only one reactant is ionized and a second
high energy to ionize both reactants. The energy state of the reactants
at both extremes often has an effect of the measured rate constant.

Table 1: Rate Constants of Positive Ion Molecule Reactions

Formaldehyde System

No.	Reaction								$k \times 10^{-9}$ $(cm^{-3}.sec^{-1})$
1	CH_2O^+	+	CH_2O	→	CH_3O^+	+	CHO		1.64
2	CHO^+	+	CH_2O	→	CH_3O^+	+	CO		0.61
3	C^+	+	CH_2O	→	CH_2^+	+	CO		0.82
4	CH_2^+	+	CH_2O	→	CH_3^+	+	CHO		0.64
5	CH^+	+	CH_2O	→	CH_3^+	+	CO		0.32
6	CH_2O^+	+	CH_2O	→	CH_4O^+	+	CO		0.13
7	CH_3O^+	+	CH_2O	→	CH_5O^+	+	CO		0.012
8					$C_2H_5O_2^+$				0.024
9	CH_3O^+	+	CH_2O		$C_2H_5O^+$	+	0		0.003_6
11					$C_2H_3O^+$	+	H_2O		0.007
12					$C_2H_2O^+$	+	H_2O	+ H	0.02

Reference reaction: $CH_4^+ + CH_4 \rightarrow CH_5^+ + CH_3$; $k = 1.2 \times 10^9$ $cm^{-3} s^{-1}$.

Table 2: Rate Constants of Positive Ion Molecule Reactions
Formyl Fluoride System.

No.	Reaction						$k \times 10^{-9}$ $(cm^{-3}.sec^{-1})$
15	$HFCO^+$	+	$HFCO$	$\rightarrow$	$H_2F_2CO^+$	+ FCO	2.5
16	HCO^+	+	$HFCO$	$\rightarrow$	H_2FCO^+	+ CO	0.71
17	CF^+	+	$HFCO$	$\rightarrow$	HFC^+	+ FCO	0.05
18	H_2FCO	+	$HFCO$	$\rightarrow$	$H_3F_2CO^+$	+ CO	0.12
19					$H_3F_2C_2O_2^+$		0.02
20	H_2FCO^+	+ $HFCO$			$H_2F_2C_2O_2^+$	+ H	0.008
21					$H_3FC_2O_2^+$	+ F	0.035
22					$H_3F_2C_2O^+$	+ O	0.05

Reference reaction: CH_4^+ + CH_4 → CH_5^+ + CH_3; 1.2×10^9 cm^{-3} sec^{-1}, 3

Table 3: Electron Energies for Ion Production by Dissociative Electron
Attachment in Phosgene, Carbonyl chlorofluoride and Carbonyl
Fluoride. Energies in Electron Volt.

Ion	Cl_2CO	$ClFCO$	F_2CO
Cl^-	th[*]; 4.0; 8.4	1.7	
Cl_2^-	th; 4.1; 8.4		
$ClCO^-$	th; 8.1	1.9	
ClF^-		1.85	
FCO^-		2.1	3.3
F_2^-			3.1
F^-		2.4	2.7[**]

[*]   virtually thermal energy.

[**]   K.A.G. Mac Neil and J.C.J. Thynne, Int. J. Mass Spectrom. Ion
Phys., 3, 35 (1969).

Table 4: Rate Constants of Negative Ion Molecule Reactions

A. Phosgene.

B. Carbonyl Fluoride.

C. Carbonyl Chlorofluoride.

No.	Reaction							$k \times 10^{-11}$ $(cm^{-3} . sec^{-1})$
A.								
23	$ClCO^-$	+	Cl_2CO	$\rightarrow$	Cl_3CO^-	+	CO	1.11
24	Cl_2^-	+	Cl_2CO	$\rightarrow$	Cl_3CO^-	+	CO	0.14
25	$ClCO^-$	+	Cl_2CO	$\rightarrow$	Cl^-	+	$Cl_2CO + Cl$	0.14
26	Cl_2^-	+	Cl_2CO	$\rightarrow$	Cl^-	+	$Cl_2CO + Cl$	0.07
B.								
27	FCO^-	+	F_2CO	$\rightarrow$	F_3CO^-	+	CO	5.8
28	F_2^-	+	F_2CO	$\rightarrow$	F_3CO^-	+	F	0.6
29	F_3CO^-	+	F_2CO	$\rightarrow$	F_5CO^-	+	CO	0.3
C.								
30	FCO^-	+	$FClCO$	$\rightarrow$	F_2ClCO^-	+	CO	1.0
31	$ClCO^-$	+	$FClCO$	$\rightarrow$	FCl_2CO^-	+	CO	3.1
32	FCl^-	+	$FClCO$	$\rightarrow$	FCl_2CO^-	+	F	0.2

Table 5: Limits for the Heat of Formation of the Negative Ions $(F, Cl)_3CO^-$

Ion	Reaction not observed	$\Delta H_{max}(kcal.mol^{-1})$	Reaction observed	$\Delta H_{max}(kcal.mol^{-1})$
Cl_3CO^-	$ClF^- + Cl_2CO \rightarrow Cl_3CO^- + F$	-108	$ClCO^- + Cl_2CO \rightarrow Cl_3CO^- + CO$	-111
FCl_2CO^-	$FCO^- + Cl_2CO \rightarrow FCl_2CO^- + CO$	-130	$ClF^- + FClCO \rightarrow FCl_2CO^- + F$	-158
F_2ClCO^-	$ClF^- + FClCO \rightarrow F_2ClCO^- + Cl$	-168	$FCO^- + FClCO \rightarrow F_2ClCO^- + CO$	-180
F_3CO^-	$ClF^- + F_2CO \rightarrow F_3CO^- + Cl$	-218	$FCO^- + F_2CO \rightarrow F_3CO^- + CO$	-230

References

1. Z. Karpas and F. S. Klein, Int. J. Mass Spectrom. Ion Phys.
   16, 289 (1975)

2. V. G. Anicich and M. T. Bowers, ibid. 11, 329 (1973)

3. T. B. McMahon and J. L. Beauchamp, Rev. Sci. Instrum. 42, 509 (1971)

4. M. Henchman, Ion-Molecule Reactions, Edit. J. L. Franklin, Plenum
   Press, New York, 1972, pp 101 - 260

5. Z. Karpas and F. S. Klein, Int. J. Mass Spectrom. Ion Phys.
   in press

6. Z. Karpas and F. S. Klein, ibid. 22, 189 (1976)

7. Z. Karpas and F. S. Klein, ibid. 18, 65 (1975)

8. G. A. Olah and S. J. Kuhn, J. Am. Chem. Soc. 82, 2380 (1960)

9. G. C. Goode, R. M. O'Malley, A.-J. Ferrer-Correia, R. I. Massey,
   K. R. Jennings, J. H. Futrell and P. M. Llewellyn, Int. J. Mass
   Spectrom. Ion Phys. 5, 393 (1970)

10. Z. Karpas and F. S. Klein, in Advances in Mass Spectrometry, vol. 7,
    Edit. N. R. Daly, Heyden, London, 1977

11. Z. Karpas, Ph. D. Thesis, Weizmann Institute, Rehovot, July 1976

Ion Chemistry of $(CH_3)_3PCH_2$, $(CH_3)_3PNH$, $(CH_3)_3PNCH_3$ and $(CH_3)_3PO$

O.-R. Hartmann, K.-P. Wanczek and H. Hartmann
Institut für physikalische Chemie der
Universität Frankfurt/Main

Introduction

Among the most important groups of organic phosphorus compounds are derivatives of phosphine and phosphoranes. Among these, phosphoranes with coordination number four are especially interesting. The ylids $(CH_3)_3PCH_2$, $(CH_3)_3PNH$, and $(CH_3)_3PO$ are isoelectronic compounds. The electronegativity of the basic center increases in the order $-CH_2$, $-PNH$, $-PO$.

Trimethylmethylenephosphorane can be described by the two structures:

$$\equiv P = CH_2 \qquad\qquad \equiv \overset{+}{P} - \bar{C}H_2^-$$

The ionic structure is the more important. The nucleophilic and electrophilic reactions of the compound indicate that the phosphorus atom is strongly electrophilic, comparable with the phosphonium atom in phosphonium salts $PR_4^+ \, X^-$. Contrary, the carbon atom is strongly nucleophilic and shows reactions characteristic for a carbanion center.

The phosphorus atom can be described with good approximation as tetraedrical sp^3 hybridized, single bonded with the planar sp^2 hybridized carbon atom. Additionally, the P-C bond has some d_π-p_π character. The bond order of the P-C bond is 1.65[1]. ^{1}H-NMR investigations show that the addition of small amounts of acids or increase of temperature[2] results in rapid proton exchange between methyl and methylene groups. Therefore each carbon atom participates in the carbanion function. Methylene transfers are important reactions in the chemistry of the ylids, e. g. the Wittig-reaction with carbonyl compounds.

Iminophosphoranes contain a NH or a NR group instead of the methylene group at the phosphorus. The reactions of the iminophosphoranes and the ylids are often analogous. Iminophosphoranes are reactive compounds, however more stable than carbon ylids. The reactivity results from the polarity of the P-N double bond and is controlled by the substituents at the phosphorus and nitrogen atoms.

$$\geqq P = N - \qquad\qquad \geqq P^+ - \bar{N}^- -$$

Substituents at the nitrogen atom with a negative inductive effect, reducing the charge at the nitrogen atom lead to a more stable and less reactive phosphine imine.[3] The corresponding is valid for the substituents at the phosphorus atom. The reactions of phosphine imines prove the dipolar structure with a nucleophilic nitrogen and an electrophilic phosphorus atom, however, the polarity is smaller than in the case of carbon ylids.

Wang[4] describes the trimethylphosphine oxide by a nearly ideal sp[3] hybridization of the phosphorus atom, according to expectations. The P-O bond can be described as a covalent bond which is superimposed by a ionic bond. The chemistry of trimethylphosphine oxide is dominated by the polar PO group.

In this work the mass spectra and ion chemistries of trimethylmethylenephosphorane, trimethyliminophosphorane, N-methyl-trimethyliminophosphorane and trimethylphosphine oxide are presented.

Experimental

The measurements were carried out with a "Syrotron" ICR spectrometer (Varian Ass. Palo Alto, California) and has been described earlier[5] in the mass range m/e 1 - 186 and at pressures not exceeding $8 \cdot 10^{-5}$torr.

Mass spectra of $(CH_3)_3PCH_2$ were also recorded with a Varian MAT mass

spectrometer CH_4 (Varian MAT,Bremen). The electron energy was 30 eV

unless otherwise stated.

All the ions which appear in the spectrum were subject to double

resonance experiments at several pressures and irradiating field

strengths. The double resonance results were supported by pressure plots

if possible. The compounds could all be handled without any decomposi-

tion in the inlet system and the measuring cell of the spectrometer.

Thermal decomposition at the hot filament was not observed. The compounds

were prepared according to literature procedures.

The mass spectrum of $(CH_3)_3PCH_2$ was recorded with a MAT CH 4 mass

spectrometer which was equipped with a room temperature ion source.

Mass Spectra

In Figures 1 - 4 the ICR mass spectra of

$(CH_3)_3PCH_2$ $(CH_3)_3PNH$ $(CH_3)_3PNCH_3$ $(CH_3)_3PO$

are depicted. Caused by the long ion paths and the long residence time

in the ICR cell there are product ions in the spectra even at pressures

of $5.6 \cdot 10^{-7}$torr from reactions with high rate constants. The appearance

of the protonated molecules underlines the high basicity of the neutral

molecules.

Comparing the ICR mass spectrum of trimethylmethylenephosphorane with

the mass spectrum recorded with the CH 4 mass spectrometer (see Table 1)

reveals interesting differences in the relative intensities of the frag-

ment ions, which are caused by the longer time interval between genera-

tion and detection of the ions in the ICR cell compared to the CH 4 mass

Table 1. ICR-Mass Spectrum and Mass Spectrum of $(CH_3)_3PCH_2$

m/e	Ion	ICR-Mass Spectrum Rel. Intensity	Mass Spectrum Rel. Intensity
14	$CH_2^{\cdot+}$		31
15	CH_3^+	31	496
16	$CH_4^{\cdot+}$	41	686
27	$C_2H_3^+$		62
28	$C_2H_4^{\cdot+}$		62
29	$C_2H_5^+$		34
32	$PH^{\cdot+}$		18
41	$C_3H_5^+$	455	467
42	$C_3H_6^{\cdot+}$		17
43	PC^+		10
44	$PCH^{\cdot+}$		48
45	PCH_2^+	249	155
46	$PCH_3^{\cdot+}$	74	42
47	$CH_2PH_2^+$	605	472
48	$H_2PCH_3^{\cdot+}$		17
49	$H_3PCH_3^+$		31
57	$P(CH)_2^+$	138	73
58	$P(CH)_2H^{\cdot+}$	61	38
59	$P(CH_2)_2^+$	300	240
60	$P(CH_2)_2H^{\cdot+}$	24	22
61	$P(CH_3)_2^+$	104	212
62	$HP(CH_3)_2^{\cdot+}$	36	27
71	$PC_3H_4^+$	63	41
73	$PC_3H_6^+$	342	257
74	$PC_3H_7^{\cdot+}$	23	26
75	$(CH_3)_2PCH_2^+$	772	936
76	$(CH_3)_3P^{\cdot+}$	93	99
77	$(CH_3)_3PH^+$	56	157
89	$(CH_3)_3PCH^+$		16
90	$(CH_3)_3PCH_2^{\cdot+}$	1000	1000
91		107	55
92		20	82

spectrometer. The proportion of ions with low m/e ratios at the total
ion intensity is higher in the ICR spectrum than in the mass spectrum,
contrary the proportion of ions with high m/e ratios* is lower. This may
indicate slow fragmentation processes which are not completed in the

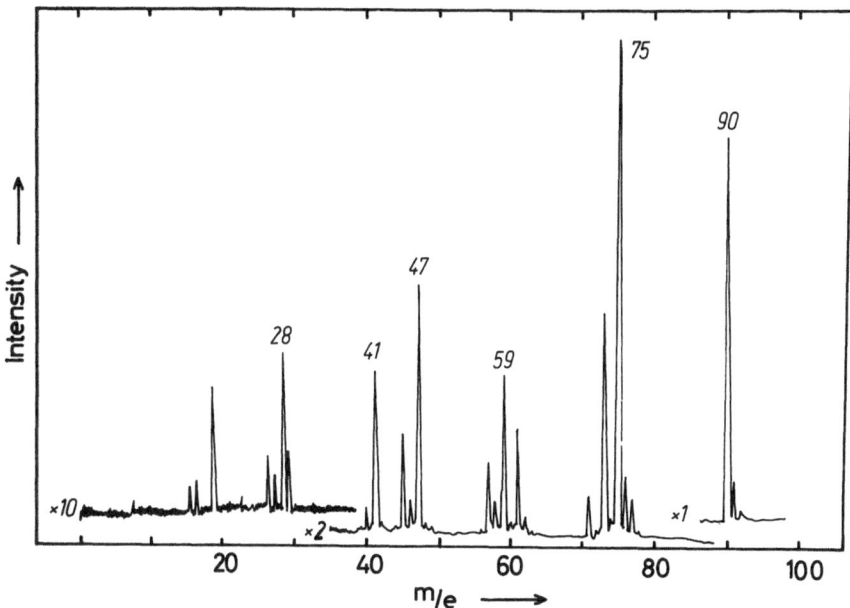

Fig. 1. ICR mass spectrum of trimethylmethylenephosphorane at a pressure
of $5.6 \cdot 10^{-7}$ torr.

mass spectrometer when the ions are detected. The intensities of the ions
m/e 75, 73, 47, 45 and 61, 59, 57 may serve as an illustration of this
effect, which needs further investigation. Their relative intensities are
compared in Table 1. In the mass spectrometer the following strong metastable decompositions were detected:

$$(CH_3)_3 PCH_2^{\cdot +} \longrightarrow PC_3H_8^+ + CH_3^{\cdot}$$

$$PC_3H_8^+ \rightleftharpoons \begin{array}{l} H_2PCH_2^+ \quad + \quad C_2H_4 \\ C_3H_5^+ \quad + \quad PH_3 \end{array}$$

These decompositions compare well with the metastable decompositions detected by Gillis and Long[6] for trimethylphosphine. In both decompositions
the ion m/e 75 plays a key role. It is formed in the case of trimethylphosphine by dissociation of a hydrogen atom, in the case of trimethylmethylenephosphorane by dissociation of a methyl group respectively from
the molecular ions.

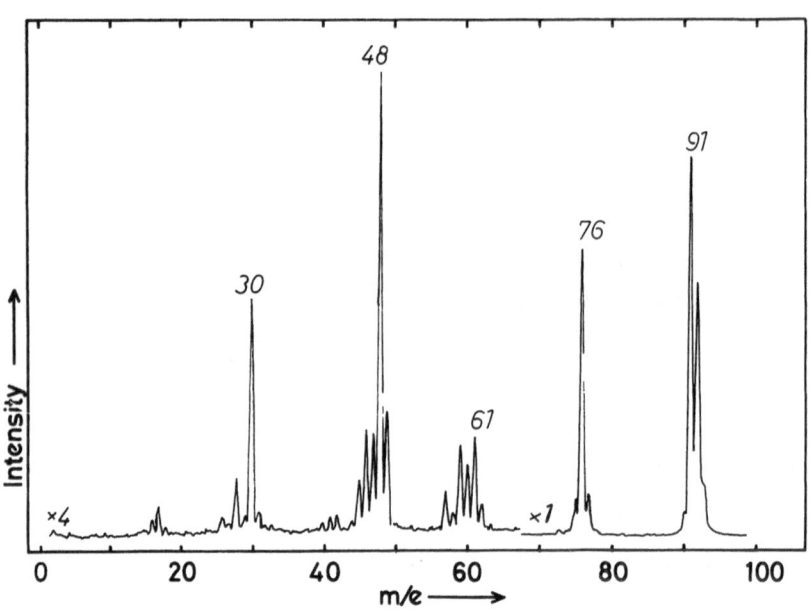

Fig. 2. ICR spectrum of trimethylphosphine imine at a pressure of
3.2·10^{-6} torr.

The molecular ions are the most abundant ions in the ICR mass spectra
of $(CH_3)_3PCH_2$, $(CH_3)_3PNH$ and $(CH_3)_3PNCH_3$, they are radical cations. All
other abundant ions, including $(CH_3)_3PO$ are even electron ions. The base

peak of trimethylphosphine oxide is the ion m/e 77, $(CH_3)_2PO^+$, formed by dissociation of a methyl radical from the molecular ion. ($M - CH_3^{\cdot}$)$^+$ fragment ions are also formed in the spectra of the ylids : $(CH_3)_3PCH_2$, $(CH_3)_3PNH$ and $(CH_3)_3PNCH_3$. In the case of $(CH_3)_3PNCH_3$ the dissociation of a CH_3NH radical is the more favoured process, the fragment ion $(CH_3)_2PCH_2^+$ is formed.

$$(CH_3)_3PNCH_3^{\cdot +} \longrightarrow (CH_3)_2PCH_2^+ + CH_3NH^{\cdot}$$

Obviously, the CH_3NH radical is stabilized by the methyl group. The corresponding $NH_2^{\cdot}$ dissociation is only a minor process. The ion m/e 75 is only low abundant in the mass spectrum of the homologous trimethylphosphine imine.

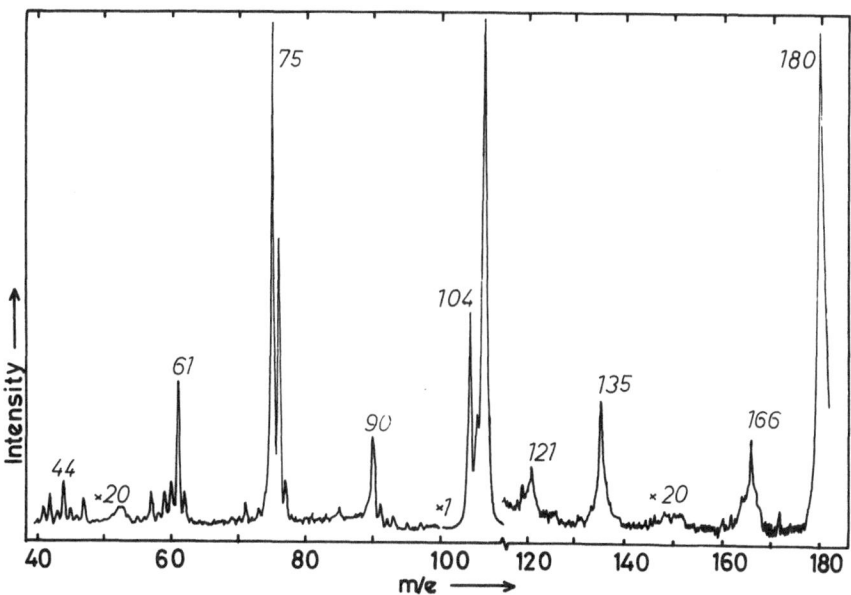

Fig. 3. ICR spectrum of N-methyl-trimethylphosphinimine at a pressure of $4 \cdot 10^{-5}$ torr.

Comparison of the mass spectra of $(CH_3)_3PNCH_3$ and the isomeric dimethylaminodimethylphosphine, $(CH_3)_2PN(CH_3)_2$, shows that in the second case also a methyl group is dissociated and a stable phosphonium fragment ion with the structures

$$(CH_3)_2PNCH_3^+ \qquad \text{or} \qquad (CH_3)PN(CH_3)_2^+$$

is formed. However, dissociation of $CH_3NH\cdot$which dominates in the fragmentation of the imine occurs only with low abundance, as is indicated by the intensity of the ion m/e 75^7.

The ion $(M - 1)^+$ is formed only in the mass spectrum of N-methyl-tri-methylphosphinimine as a medium abundant ion and not in trimethylphosphin-imine. This indicates a resonance-stabilized structure of the ion:

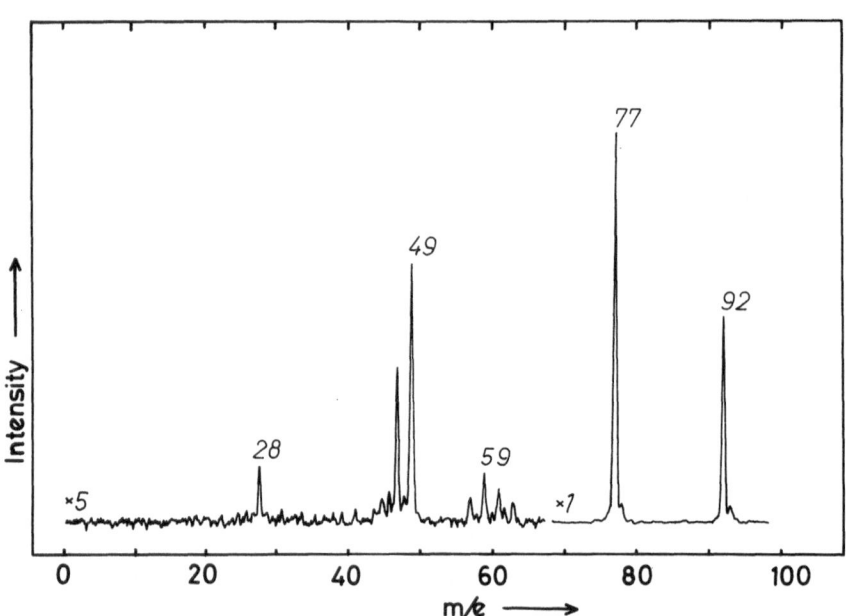

Fig. 4. ICR mass spectrum of trimethylphosphine oxide at a pressure of $1.2 \cdot 10^{-7}$ torr.

$$(CH_3)_3P=\underline{N}-CH_2^+ \qquad \longleftrightarrow \qquad (CH_3)_3P^{\pm}\underline{N}=CH_2$$

All the phosphonium ions with two methyl groups formed from the ylids $(CH_3)_2P^+CH_2$, $(CH_3)_2P^+NH$, $(CH_3)_2P^+NCH_3$, $(CH_3)_2P^+O$ can decompose further by dissociation of neutral ethene molecule:

$$(CH_3)_2P^+CH_2 \qquad \longrightarrow \qquad H_2PCH_2^+ \;+\; C_2H_4$$

$$(CH_3)_2P^+NH \qquad \longrightarrow \qquad H_2PNH^+ \;+\; C_2H_4$$

$$(CH_3)_2P^+NCH_3 \qquad \longrightarrow \qquad H_2PNCH_3^+ \;+\; C_2H_4$$

$$(CH_3)_2P^+O \qquad \longrightarrow \qquad H_2PO^+ \;+\; C_2H_4$$

The methyl groups bonded to phosphorus are replaced by hydrogen atoms.

The ion $H_2PCH_2^+$ corresponds to the ion $H_2NCH_2^+$ which is observed frequently in the mass spectra of alkyl amines. Hydrogen molecules are dissociated also in the formation of fragment ions, as is indicated by groups of ions like $(CH_3)_2P^+$, $(CH_2)_2P^+$ and $(CH)_2P^+$.

Ion-Molecule Reactions

Trimethylmethylenephosphorane

In Figure 5 an ICR spectrum of trimethylmethylenephosphorane at a pressure of $1.6 \cdot 10^{-5}$ torr is shown. With the exception of the ion m/e 104, the abundant product ions have even electron numbers. Numerous product ions show m/e ratios which are different by 90 mass units from the masses of abundant fragment ions. However, double resonance experiments indicate that those ions are not formed by addition reactions of the respective fragment ions to the neutral molecule (the neutral molecule has a molecular weight of 90 mass units). Only the molecular ion and the heavy fragment ions undergo the addition reactions to be expected.

In Table 2 all ion-molecule reactions of trimethylmethylenephosphorane are listed. The structures of the ions are discussed below, they

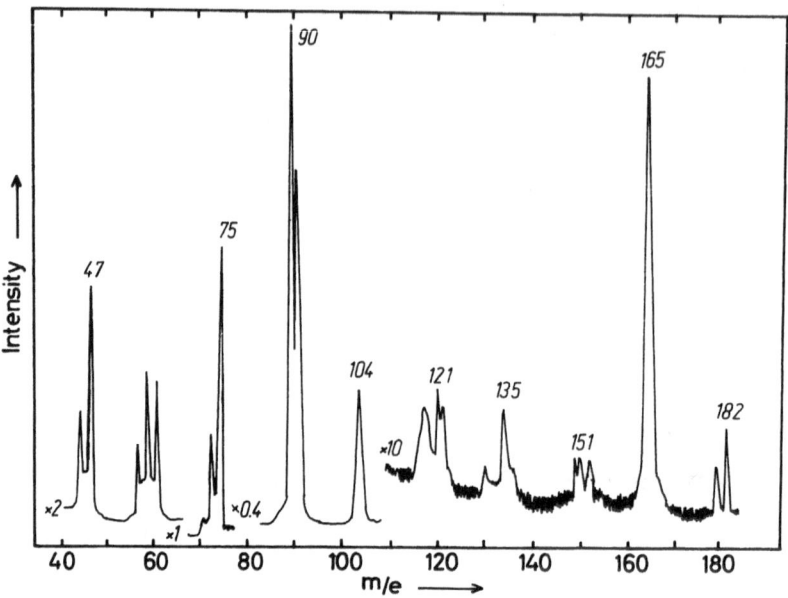

Fig. 5. ICR spectrum of trimethylmethylenephosphorane at a pressure of
$1.6 \cdot 10^{-5}$ torr.

are deduced from the ion chemistry and by analogizing with the chemistry
of the condensed phase.

The molecular ion undergoes numerous ion-molecule reactions. Its
structure

$$(CH_3)_3 \overset{+}{P} \overset{-}{-} CH_2^{\bullet}$$

is deduced from PE and ESR measurements. The first band in the PE spectrum
of trimethylmethylenephosphorane [8] observed at 6.9 eV is due to dissociation of an electron from the carbanion site of the molecule. According to
calculations by Absar et al.[9] and Hoffmann et al.[10] the electron pair
with highest energy is located almost completely at the carbon atom. ESR
measurements on the radical cation which has been generated from triphe-

nylmethylenephosphorane by X-rays show that this cation has the structure

$$(C_6H_5)_3\overset{+}{P}-\overset{\cdot}{C}H_2$$

As a first step in the ion-molecule reactions of the molecular ion with a neutral trimethylmethylenephosphorane molecule addition with formation of a four center intermediate is assumed. An analogous intermediate is observed in the Wittig reaction[3].

$$(CH_3)_3P\overset{\cdot+}{C}H_2 + (CH_3)_3PCH_2 \longrightarrow (CH_3)_3\underset{\underset{CH_2-P(CH_3)_3}{|}}{P}-\overset{\cdot+*}{\underset{|}{C}H_2} \longrightarrow products$$

The dimer molecular ion can be detected as a low abundant stable product ion.

$$(CH_3)_3P\overset{\cdot+}{C}H_2 + (CH_3)_3PCH_2 \longrightarrow ((CH_3)_3PCH_2)\overset{\cdot+}{2} \qquad (2.43)$$

In a second step of the reaction the intermediate complex dissociates to form product ions and neutral particles.

The ions formed in the reactions (2.23) m/e 121, (2.26) m/e 122, (2.27) m/e 123, (2.31) m/e 135, (2.33) m/e 136, (2.38) m/e 150, (2.39) m/e 151 and (2.42) m/e 165 can be described by the general formula $(CH_3)_3PCHPR_n^+$ ($R = CH_3$, H; n = 1,2,3).

As an example the formation of the ion m/e 165, trimethyltrimethylphosphoniummethylenephosphorane, may be discussed.

$$(CH_3)_3P\overset{\cdot+}{C}H_2 + (CH_3)_3PCH_2 \longrightarrow (CH_3)_3PCHP(CH_3)_3^+ + \overset{\cdot}{C}H_3 \qquad (2.42)$$

This symmetrical ion is stabilized by resonance:

$$(CH_3)_3P=CH-\overset{+}{P}(CH_3)_3 \rightleftharpoons (CH_3)_3\overset{+}{P}-CH=P(CH_3)_3$$

Trimethyltrimethylphosphoniummethylenephosphorane chloride is a stable compound which has been prepared[11] according to the following reactions:

$$(CH_3)_3PCH_2 + (CH_3)_2PCl \longrightarrow (CH_3)_3PCH_2P(CH_3)_2^+ \ Cl^-$$

$$(CH_3)_3PCH_2P(CH_3)_2^+ \ Cl^- + (CH_3)_3PCH_2 \longrightarrow (CH_3)_3PCHP(CH_3)_2 + (CH_3)_4P^+ \ Cl^-$$

$$(CH_3)_3PCHP(CH_3)_2 + CH_3J \longrightarrow (CH_3)_3PCHP(CH_3)_3^+ \ J^-$$

The product ion $(CH_3)_3PCHP(CH_3)_3^+$ is formed by an addition reaction of

the fragment ion m/e 75 (2.41) and a reaction of the molecular ion with

the neutral ylid molecule.

$$(CH_3)_2PCH_2^+ + (CH_3)_3PCH_2 \longrightarrow (CH_3)_3PCHP(CH_3)_3^+ \qquad (2.41)$$

Eight of the nine product ions possible with the general formula

$(CH_3)_3PCHPR_n^+$ (R = CH_3, H; n = 1,2,3) are detected. Only the ion m/e

137, $(CH_3)_3PCHP(CH_3)H_2^+$ is not formed by this group of reactions. All

the products discussed until now contain this structure with great pro-

bability because it is stabilized by resonance. In many of the reac-

tions rearrangements must be assumed, as indicated by the formation of

the P-H bonds. Such rearrangements are also known in the chemistry of

the ylids in the condensed phase[11].

An intense ion-molecule reaction of the molecular ion which does not

yield a product of the general formula $(CH_3)_3PCHPR_n^+$ is the methylene

transfer reaction (2.16).

$$(CH_3)_3PCH_2^{\cdot+} + (CH_3)_3PCH_2 \longrightarrow (CH_3)_3P(CH_2)_2^{\cdot+} + (CH_3)_3P \quad (2.16)$$

Methylene transfer reactions are a second important class of reactions

of trimethylmethylenephosphorane[3]. In the ion chemistry a neutral ylid

molecule transfers a methylene group to an ion and a neutral trimethyl-

phosphine molecule is formed. A second channel is observed for this re-

action where the position of the charge is changed between the products.

$$(CH_3)_3PCH_2^{\cdot+} + (CH_3)_3PCH_2 \longrightarrow (CH_3)_3P^{\cdot+} + PC_5H_{13} \qquad (2.3)$$

In the second reaction trimethylphosphine is the ionic product of the

reaction.

Table 2. Ion-Molecule Reactions of Trimethylmethylenephosphorane

m/e	Product Ion	m/e	Primary Ion	No. of Reaction
61	$P(CH_3)_2^+$	47	$H_2PCH_2^+$	(2.1)
76	$P(CH_3)_3^{\cdot +}$	47	$H_2PCH_2^+$	(2.2)
		90	$(CH_3)_3PCH_2^{\cdot +}$	(2.3)
		104	$(CH_3)_2(C_2H_5)PCH_2^{\cdot +}$	(2.4)
77	$P(CH_3)_3^{\cdot +}$	47	$H_2PCH_2^+$	(2.5)
87	$PC_4H_8^+$	73	$PC_3H_6^+$	(2.6)
		75	$(CH_3)_2PCH_2^+$	(2.7)
91	$P(CH_3)_4^+$	47	$H_2PCH_2^+$	(2.8)
		75	$(CH_3)_2PCH_2^+$	(2.9)
		90	$(CH_3)_3PCH_2^{\cdot +}$	(2.10)
		104	$(CH_3)_2(C_2H_5)PCH_2^{\cdot +}$	(2.11)
92	$P_2C_2H_6^{\cdot +}$	77	$HP(CH_3)_3^+$	(2.12)
103	$PC_5H_{12}^+$	75	$(CH_3)_2PCH_2^+$	(2.13)
		91	$(CH_3)_4P^+$	(2.14)
104	$(CH_3)_2(C_2H_5)PCH_2^{\cdot +}$	75	$(CH_3)_2PCH_2^+$	(2.15)
		90	$(CH_3)_3PCH_2^{\cdot +}$	(2.16)
105	$(CH_3)_3(C_2H_5)P^+$	91	$(CH_3)_4P^+$	(2.17)
106	$P_2C_3H_8^{\cdot +}$	91	$(CH_3)_4P^+$	(2.18)
107	$P_2C_3H_9^+$	91	$(CH_3)_4P^+$	(2.19)
118	$(CH_3)(C_2H_5)_2PCH_2^{\cdot +}$	104	$(CH_3)_2(C_2H_5)PCH_2^{\cdot +}$	(2.20)
119	$(CH_3)_2(C_2H_5)_2P^+$	105	$(CH_3)_3(C_2H_5)P^+$	(2.21)
121	$(CH_3)_3PCHPH^+$	73	$PC_3H_6^+$	(2.22)
		90	$(CH_3)_3PCH_2^{\cdot +}$	(2.23)
		107	$P_2C_3H_9^+$	(2.24)
122	$(CH_3)_3PCHPH_2^{\cdot +}$	73	$PC_3H_6^+$	(2.25)
		90	$(CH_3)_3PCH_2^{\cdot +}$	(2.26)
123	$(CH_3)_3PCHPH_3^+$	90	$(CH_3)_3PCH_2^{\cdot +}$	(2.27)
131	$C_3H_5CH_2P(CH_3)_3^+$	41	$C_3H_5^+$	(2.28)
		91	$(CH_3)_4P^+$	(2.29)
135	$(CH_3)_3PCHP(CH_3)^+$	75	$(CH_3)_2PCH_2^+$	(2.30)
		90	$(CH_3)_3PCH_2^{\cdot +}$	(2.31)
		121	$(CH_3)_3PCHPH^+$	(2.32)
136	$(CH_3)_3PCHP(CH_3)H^{\cdot +}$	90	$(CH_3)_3PCH_2^+$	(2.33)

Table 2 (cont'd)

m/e	Product Ion	m/e	Primary Ion	No. of Reaction
137	$(CH_3)_3PCHP(CH_3)H_2^+$	75	$(CH_3)_2PCH_2^+$	(2.34)
147	$(CH_3)_3PCHP(CH)_2H^+$	90	$(CH_3)_3PCH_2^{\cdot+}$	(2.35)
149	$(CH_3)_3PCHP(CH_2)_2H^+$	90	$(CH_3)_3PCH_2^{\cdot+}$	(2.36)
150	$(CH_3)_3PCHP(CH_3)_2^{\cdot+}$	75	$(CH_3)_2PCH_2^+$	(2.37)
		90	$(CH_3)_3PCH_2^{\cdot+}$	(2.38)
151	$(CH_3)_3PCHP(CH_3)_2H^+$	90	$(CH_3)_3PCH_2^{\cdot+}$	(2.39)
		121	$(CH_3)_3PCHPH^+$	(2.40)
165	$(CH_3)_3PCHP(CH_3)_3^+$	75	$(CH_3)_2PCH_2^+$	(2.41)
		90	$(CH_3)_3PCH_2^{\cdot+}$	(2.42)
180	$((CH_3)_3PCH_2)_2^{\cdot+}$	90	$(CH_3)_3PCH_2^{\cdot+}$	(2.43)

The ion m/e 104 formed in reaction (2.16) rearranges with great pro-

bability:

$$(CH_3)_3P \underset{CH_2^{\cdot}}{\overset{CH_2^+}{<}} \longrightarrow (CH_3)_2(C_2H_5)PCH_2^{\cdot+}$$

to a phosphonium radical ion which resembles the molecular ion, because

the ion m/e 104 undergoes a further methylene transfer reaction to the

product ion m/e 118

$$(CH_3)_2(C_2H_5)PCH_2^{\cdot+} + (CH_3)_3PCH_2 \rightarrow (CH_3)(C_2H_5)_2PCH_2^{\cdot+} + (CH_3)_3P \quad (2.20)$$

The ion m/e 104 also transfers a proton to the neutral molecule, reac-

tion (2.11).

A further sequence of methylene transfer reactions starts with a

reaction of the protonated molecule

$$(CH_3)_4P^+ + (CH_3)_3PCH_2 \longrightarrow (CH_3)_3PC_2H_5^+ + (CH_3)_3P \quad (2.17)$$

$$(CH_3)_3PC_2H_5^+ + (CH_3)_3PCH_2 \longrightarrow (CH_3)_2P(C_2H_5)_2^+ + (CH_3)_3P \quad (2.21)$$

It leads to the tertiary ion m/e 105 and the quarternary ion m/e 119.

These ions are also assumed to rearrange to the more stable phosphonium ions. It should be mentioned that in the condensed phase only a few pen-taalkylphosphoranes are known.

Four further methylene transfer reactions are detected for the ions m/e 47 (2.1), 73 (2.6), 107 (2.24) and 121 (2.32). In the reaction (2.1) the product ion m/e 61 is formed. This ion has the structure $(CH_3)_2P^+$ if the rearrangement discussed earlier is assumed.

$$H_2PCH_2^+ \quad + \quad (CH_3)_3PCH_2 \quad \longrightarrow \quad (CH_3)_2P^+ \quad + \quad (CH_3)_3P \qquad (2.1)$$

The most intense fragment ion m/e 75 is formed by dissociation of a methyl radical from the molecular ion. Two structures may be assigned to this ion according to its reactions:

$$(a) \quad \begin{array}{c} CH_3 \\ \diagdown \\ \diagup \\ CH_3 \end{array} P^+ \!\!=\!\! CH_2 \qquad (b) \quad \begin{array}{c} CH_2 \\ | \\ CH_2 \end{array}\!\!\!\!\diagdown\!\!\!\diagup P^+ \begin{array}{c} CH_3 \\ \diagdown \\ H \end{array}$$

Most of the reactions proceed to product ions with the general formula $(CH_3)_3PCHPR_n^+$ discussed already for the molecular ion. Therefore the structure (a) which resembles that of the molecular ion is assumed. However the ion m/e 75 also reacts to the ion m/e 137 which has not been observed in the ion chemistry of the molecular ion.

$$(CH_2)_2P(CH_3)H^+ \quad + \quad (CH_3)_3PCH_2 \quad \longrightarrow \quad (CH_3)_3PCHP(CH_3)H_2^+ \quad + \quad C_2H_4 \quad (2.34)$$

Reaction (2.34) is formally a CH_3PH^+ ion transfer. It is well known from the ion chemistry of the three-membered ring compound phosphirane[12] that ring compounds transfer a small group or atom with simultaneous forma-tion of a stable neutral particle. For example, the molecular ion of ND_3 reacts with the neutral phosphirane molecule to form $HPND_3^{+\cdot}$ and $C_2H_4\cdot$. Reaction (2.34) may be interpreted in the same manner, indicating the cyclic structure (b) for the ion m/e 75.

The product ion m/e 135 is formed by three reactions.

$$(CH_3)_2PCH_2^+ + (CH_3)_3PCH_2 \longrightarrow (CH_3)_3PCHPCH_3^+ + C_2H_6 \qquad (2.30)$$

$$(CH_3)_3PCH_2^{\cdot +} + (CH_3)_3PCH_2 \longrightarrow (CH_3)_3PCHPCH_3^+ + C_3H_9^{\cdot} \qquad (2.31)$$

$$(CH_3)_3PCHPH^+ + (CH_3)_3PCH_2 \longrightarrow (CH_3)_3PCHPCH_3^+ + (CH_3)_3P \qquad (2.32)$$

Reaction (2.32) is again a methylene transfer.

The first step of reactions (2.13) and (2.7) may be a nucleophilic attack of a neutral ylid molecule at the carbon atom of the methylene group instead of the phosphorus atom of the ion. In the first case the intermediate stabilizes by migration of a proton followed by dissociation of a neutral dimethylphosphine molecule.

$$(CH_3)_2PCH_2^+ + (CH_3)_3PCH_2 \longrightarrow (CH_3)_3PCHCH_2^+ + (CH_3)_2PH \qquad (2.13)$$

$$(CH_3)_2PCH_2^+ + (CH_3)_3PCH_2 \longrightarrow (CH_2)_2P(CH_2)_2^+ + (CH_3)_3P + H_2 \qquad (2.7)$$

Reaction (2.7) leads to the product ion m/e 87 and the neutral molecules H_2 and $(CH_3)_3P$. The product ion m/e 87 is also formed by the primary ion m/e 73. This reaction supports the structure assumed for the product ion.

$$(CH_2)_2PCH_2^+ + (CH_3)_3PCH_2 \longrightarrow (CH_2)_2P(CH_2)_2^+ + (CH_3)_3P \qquad (2.6)$$

Reaction (2.6) is again a methylene transfer reaction. Reaction (2.9), a proton transfer reaction may also confirm the structure (b) for the ion m/e 75.

$$(CH_2)_2P(CH_3)H^+ + (CH_3)_3PCH_2 \longrightarrow (CH_3)_4P^+ + (CH_2)_2PCH_3 \qquad (2.9)$$

The proton affinity of trimethylmethylenephosphorane is very likely to be larger than that of methylphosphirane. The determination of the two proton affinities is going on in this laboratory. Beside reaction (2.9) the protonated molecule is also formed by the reactions (2.8), (2.10) and (2.11).

$$H_2PCH_2^+ + (CH_3)_3PCH_2 \longrightarrow (CH_3)_4P^+ + PCH_3 \qquad (2.8)$$

$$(CH_3)_3PCH_2^{\cdot +} + (CH_3)_3PCH_2 \longrightarrow (CH_3)_4P^+ + PC_4H_{10}^{\cdot} \qquad (2.10)$$

$$(CH_3)_2(C_2H_5)PCH_2^{\cdot +} + (CH_3)_3PCH_2 \longrightarrow (CH_3)_4^+ + PC_5H_{12}^{\cdot} \qquad (2.11)$$

The fragment ion m/e 41 has the composition $C_3H_5^+$. It is formed by a metastable decomposition of the ion m/e 75 in the case of trimethylmethylenephosphorane and trimethylphosphine. During a strong rearrangement all P-C bonds have been cleaved and new C-C bonds have been formed. The ion undergoes an addition reaction with trimethylphosphine[13] and trimethylmethylenephosphorane (2.28).

$$C_3H_5^+ + (CH_3)_3PCH_2 \longrightarrow (CH_3)_3PCH_2C_3H_5^+ \qquad (2.28)$$

The reaction (2.29) of the protonated molecule leads to the product ion m/e 131, too.

$$(CH_3)_4P^+ + (CH_3)_3PCH_2 \longrightarrow (CH_3)_3PCH_2C_3H_5^+ + CH_3PH_2 + 2H_2 \quad (2.29)$$

This ion is not formed by a reaction of the molecular ion. Therefore dissociation of neutral CH_3PH_2 and H_2 molecules in the course of reaction (2.29) can be assumed. Such a dissociation is not possible in a reaction of the molecular ion.

Trimethyliminophosphorane

The ion-molecule reactions of $(CH_3)_3PNH$ are summarized in Table 3. Reactions (3.1) to (3.14) yield product ions with the same masses as fragment ions, they may be described as collision dissociation reactions.

The molecular ion is the base peak in the spectrum. It is nearly unreactive. Only two ion-molecule reactions could be detected. The first (3.13) is a collision dissociation reaction yielding the product ion $(CH_3)_2P^+$, m/e 61. The second leads to the protonated molecule, m/e 92, which is an aminophosphonium ion.

$$(CH_3)_3PNH^{\cdot +} + (CH_3)_3PNH \longrightarrow (CH_3)_3PNH_2^+ + PNC_3H_9^{\cdot} \qquad (3.18)$$

Table 3. Ion-Molecule Reactions of Trimethyliminophosphorane

m/e	Product Ion	m/e	Primary Ion	No. of Reaction
30	$CH_2NH_2^+$	76	$(CH_3)_2PNH^+$	(3.1)
		92	$(CH_3)_3PNH_2^+$	(3.2)
46	PNH^+	48	H_2PNH^+	(3.3)
		76	$(CH_3)_2PNH^+$	(3.4)
		92	$(CH_3)_3PNH_2^+$	(3.5)
47	$H_2PCH_2^+$	92	$(CH_3)_3PNH_2^+$	(3.6)
48	H_2PNH^+	76	$(CH_3)_2PNH^+$	(3.7)
		92	$(CH_3)_3PNH_2^+$	(3.8)
59	$(CH_2)_2P^+$	92	$(CH_3)_3PNH_2^+$	(3.9)
60	$PC_2H_5^+$	76	$(CH_3)_2PNH^+$	(3.10)
		92	$(CH_3)_3PNH_2^+$	(3.11)
61	$(CH_3)_2P^+$	76	$(CH_3)_2PNH^+$	(3.12)
		91	$(CH_3)_3PNH^{\cdot +}$	(3.13)
75	$PC_3H_8^+$	46	PNH^+	(3.14)
92	$(CH_3)_3PNH_2^+$	30	$CH_2NH_2^+$	(3.15)
		48	H_2PNH^+	(3.16)
		76	$(CH_3)_2PNH^+$	(3.17)
		91	$(CH_3)_3PNH^{\cdot +}$	(3.18)
104	$(CH_3)_3PNCH_2^+$	76	$(CH_3)_2PNH^+$	(3.19)
		92	$(CH_3)_3PNH_2^+$	(3.20)
105	$(CH_3)_3PN(CH_3)^{\cdot +}$	92	$(CH_3)_3PNH_2^+$	(3.21)
106	$(CH_3)_3PN(CH_3)H^+$	92	$(CH_3)_3PNH_2^+$	(3.22)
122	$(CH_3)_3PPNH^+$	48	H_2PNH^+	(3.23)
		76	$(CH_3)_2PNH^+$	(3.24)
		92	$(CH_3)_3PNH_2^+$	(3.25)
132	$(CH_3)_3PNHC_3H_5^+$	41	$C_3H_5^+$	(3.26)
137	$(CH_3)_3PP(CH_3)_2^+$	61	$(CH_3)_2P^+$	(3.27)
		122	$(CH_3)_3PPNH^+$	(3.28)
152	$(CH_3)_3PNP(CH_3)_2H^+$	76	$(CH_3)_2PNH^+$	(3.29)
		92	$(CH_3)_3PNH_2^+$	(3.30)
167	$(CH_3)_3PNP(CH_3)_2NH_2^+$	76	$(CH_3)_2PNH^+$	(3.31)
		92	$(CH_3)_3PNH_2^+$	(3.32)
183	$(CH_3)_3PNHP(CH_3)_3NH_2^+$	92	$(CH_3)_3PNH_2^+$	(3.33)

The stable amoinophosphonium ion is also formed by reactions of the fragment ions m/e 30, 48 and 76.

$$CH_2NH_2^+ \; + \; (CH_3)_3PNH \longrightarrow (CH_3)_3PNH_2^+ \; + \; NCH_3 \qquad (3.15)$$

$$H_2PNH^+ \; + \; (CH_3)_3PNH \longrightarrow (CH_3)_3PNH_2^+ \; + \; PNH_2 \qquad (3.16)$$

$$(CH_3)_2PNH^+ + (CH_3)_3PNH \longrightarrow (CH_3)_3PNH_2^+ \; + \; PNC_2H_6 \qquad (3.17)$$

Contrary to the molecular ion, the protonated molecule is very reactive. Almost all product ions appearing in the spectrum are formed by reactions of the protonated molecule, too. They are summarized in Figure 6. As indicated in the figure the first step in the reaction is a nucleophilic attack of the neutral trimethyliminophosphorane at the phosphonium ion leading to the intermediate m/e 183. The intermediate can stabilize without dissociation of a neutral particle. The tertiary ion m/e 183 most probably has the structure shown in Figure 6 and not a proton-bridged structure.

The intermediate can also stabilize under dissociation of neutral particles. The product ion m/e 167 is generated by dissociation of CH_4. This ion is also formed in an addition reaction of the fragment ion m/e 76

$$(CH_3)_3PNH_2^+ + (CH_3)_3PNH \longrightarrow (CH_3)_3PNP(CH_3)_2NH_2^+ + CH_4 \qquad (3.32)$$

$$(CH_3)_2PNH^+ + (CH_3)_3PNH \longrightarrow (CH_3)_3PNP(CH_3)_2NH_2^+ \qquad (3.31)$$

The CH_4 dissociated in reaction (3.32) is probably composed of a methyl group of the fivefold coordinated phosphorus atom and of a hydrogen atom of the bridge amino group. The product ions of reactions (3.31) and (3.32) are assumed to rearrange to the resonance-stabilized structure $\equiv P{=}\underline{N}{-}P^{+\!-}_{\lessgtr}$ which corresponds to the $\equiv P{=}CH{-}P^{+\!-}_{\lessgtr}$ structure already discussed in the ion chemistry of $(CH_3)_3PCH_2$.

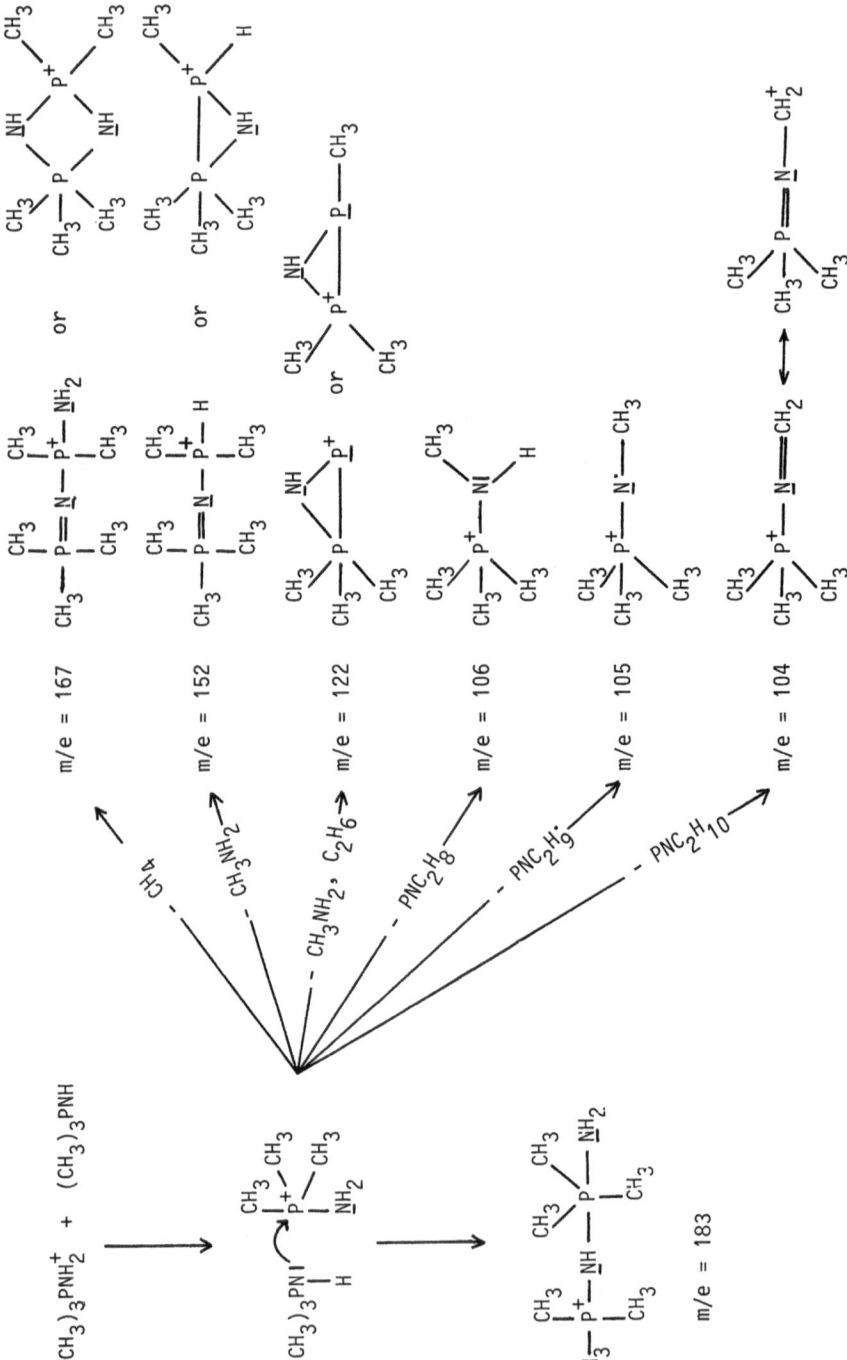

Fig. 6. Ion-molecule reactions of the protonated molecule of trimethylphosphine imine

$$H_3C-\overset{\overset{\displaystyle CH_3}{|}}{\underset{\underset{\displaystyle CH_3}{|}}{P^+}}-\bar{N}H-\overset{\overset{\displaystyle CH_3}{|}}{\underset{\underset{\displaystyle CH_3}{|}}{P}}=\bar{N}H \longrightarrow H_3C-\overset{\overset{\displaystyle CH_3}{|}}{\underset{\underset{\displaystyle CH_3}{|}}{P}}=\bar{N}-\overset{\overset{\displaystyle CH_3}{|}}{\underset{\underset{\displaystyle CH_3}{|}}{P^+}}-\bar{N}H_2$$

However, a further stabilization reaction of the ion m/e 167 may be dis-

cussed: The ion can form a ring structure via intramolecular nucleophi-

lic attack of the nitrogen atom at the phosphonium ion site:

$$\begin{array}{c} CH_3 \\ CH_3 \end{array} P \underset{\bar{N}H}{\overset{\bar{N}H}{<}} \bigg\rangle P^+ \overset{CH_3}{\underset{CH_3}{<}}$$

It has been shown by Clemens et al.[14] that resonance-stabilized cat-

ions can be formed by migration of the radical R in the course of the

reaction of chloroamine, ammonia and $(C_6H_5)_2P-\bar{N}R-P(C_6H_5)_2$.

$(C_6H_5)_2P-\bar{N}R-P(C_6H_5)_2 + 2\ NH_2Cl + NH_3 \rightarrow (H_2N(C_6H_5)_2P=\bar{N}-P(C_6H_5)_2NHR)^+\ Cl^-$
$+\ NH_4Cl$

The product ion m/e 152 is formed by the ions m/e 76 and 92.

$$(CH_3)_2PNH^+ + (CH_3)_3PNH \longrightarrow (CH_3)_3PNP(CH_3)_2H^+ + NH \qquad (3.29)$$

$$(CH_3)_3PNH_2^+ + (CH_3)_3PNH \longrightarrow (CH_3)_3PNP(CH_3)_2H^+ + CH_3NH_2 \ (3.30)$$

In reaction (3.29) neutral products with mass number 15 and in reaction

(3.30) with mass number 31 are formed. The intermediates of the reactions

have even electron numbers and this should be true also for the product

ion. Therefore the structures NH and CH_3NH_2 are the most probable for

the neutral reaction products. The product ion m/e 152 may have a cyc-

lic or an open-chained resonance-stabilized structure.

$$(CH_3)_3P=\bar{N}-P^+(CH_3)_2H$$

$$(CH_3)_3P^{\pm}\bar{N}=P(CH_3)_2H$$

$$(CH_3)_3P \underset{}{\overset{\bar{N}H}{\diagdown}} P^+(CH_3)_2$$

The cyclic structure contains a P-P bond. The structure of the ion m/e

152 is in agreement with the corresponding reaction of N-methyl-trime-thyliminophosphorane:

$$(CH_3)_3PN(CH_3)H^+ + (CH_3)_3PNCH_3 \rightarrow (CH_3)_3PNP(CH_3)_3^+ + (CH_3)_2NH \qquad (4.37)$$

In the course of this reaction $(CH_3)_2NH$ is dissociated. The ion m/e 166 will be discussed in detail later.

Also the ion m/e 122 is formed by reactions of $(CH_3)_2PNH^+$ and $(CH_3)_3PNH_2^+$:

$$(CH_3)_2PNH^+ + (CH_3)_3PNH \longrightarrow (CH_3)_3PPNH^+ + (CH_3)_2NH \qquad (3.24)$$

$$(CH_3)_3PNH_2^+ + (CH_3)_3PNH \longrightarrow (CH_3)_3PPNH^+ + CH_3NH_2 + C_2H_6 \qquad (3.25)$$

The reactions (3.24) and (3.25) are P^+ transfers to the neutral $(CH_3)_3PNH$ molecule. The ion m/e 48 undergoes a P^+ transfer reaction, too leading to the same product ion:

$$H_2PNH^+ + (CH_3)_3PNH \longrightarrow (CH_3)_3PPNH^+ + NH_3 \qquad (3.23)$$

In reaction (3.23) ammonia is formed as neutral reaction product, the neutral product in reaction (3.24) is dimethylamine, in reaction (3.25) the most probable neutral products are methylamine and ethane.

The ion m/e 122 may have structures corresponding to the structures discussed earlier:

$$(CH_3)_2P^+ = \bar{N} - P(CH_3)H$$

$$(CH_3)_3P \overset{\displaystyle \bar{N}H}{\underset{\displaystyle }{\diagup \diagdown}} P^+$$

$$(CH_3)_2P \overset{\displaystyle \bar{N}H}{\underset{\displaystyle }{\diagup \diagdown}} \overset{+}{\underline{P}}CH_3$$

The reaction of the ion m/e 122 to form the product ion m/e 137 is not understood until now:

$$(CH_3)_3PPNH^+ + (CH_3)_3PNH \longrightarrow (CH_3)_3PP(CH_3)_2^+ + PN_2CH_5 \qquad (3.28)$$

This reaction does not seem to proceed via simple CH_3^- or NH transfers. The product ion m/e 137 is formed in a second reaction by the primary ion m/e 61:

$$(CH_3)_2P^+ + (CH_3)_3PNH \longrightarrow (CH_3)_3PP(CH_3)_2^+ + NH \qquad (3.27)$$

A first step of this reaction is addition of the neutral phosphinimine to the $(CH_3)_2P^+$ ion:

$$(CH_3)_3P{=}NH \longrightarrow P^+(CH_3)_2 \longrightarrow (CH_3)_3P^+{-}\bar{N}H{-}\bar{P}(CH_3)_2$$

$$\longrightarrow (CH_3)_3P^+ \overset{}{\underset{\bar{N}H}{\diagdown}} \overset{\diagup}{P(CH_3)_2} \longrightarrow (CH_3)_3PP(CH_3)_2^+ + NH$$

followed by dissociation of an NH molecule and formation of a P-P bond in the ion. If formation of a methyl and not a NH molecule is assumed, the product ion must be a radical cation. Formation of two odd electron products is however unlikely. A corresponding reaction is detected in the case of N-methyl-trimethylphosphinimine. In this reaction a NCH_3 molecule is dissociated and a P-P bond is also formed in the ion. The ion m/e 137 is the most intense product ion in the ion chemistry of tri-methylphosphine[13]. It is well known in the chemistry of the condensed phase[16].

The primary ion m/e 41 undergoes only an addition reaction, corresponding reactions have been already discussed for $(CH_3)_3PCH_2$ and $(CH_3)_3P$:

$$C_3H_5^+ + (CH_3)_3PNH \longrightarrow (CH_3)_3PNHC_3H_5^+ \qquad (3.26)$$

The product ions m/e 104, 105 and 106 are only minor ions. Their formation reactions are

$$(CH_3)_2PNH^+ + (CH_3)_3PNH \longrightarrow (CH_3)_3PNCH_2^+ + PNCH_5 \qquad (3.19)$$

$$(CH_3)_3PNH_2^+ + (CH_3)_3PNH \longrightarrow (CH_3)_3PNCH_2^+ + PNC_2H_{10} \qquad (3.20)$$

$$(CH_3)_3PNH_2^+ + (CH_3)_3PNH \longrightarrow (CH_3)_3PNCH_3^{\cdot+} + PNC_2H_9^{\cdot} \qquad (3.21)$$

$$(CH_3)_3PNH_2^+ + (CH_3)_3PNH \longrightarrow (CH_3)_3PN(CH_3)H^+ + PNC_2H_8 \qquad (3.22)$$

In the reactions probably a nitrogen-carbon bond is formed. Reaction (3.22) yields the N-methyl-trimethylaminophosphonium ion, reaction (3.20) a resonance-stabilized product.

N-methyl-trimethyliminophosphorane

The ion-molecule reactions of N-methyl-trimethyliminophosphorane are summarized in Table 4. The reactions can again be divided into collision dissociation reactions, which yield product ions with lower masses than the molecular ion and reactions which yield product ions with greater masses. In the second case most of the neutral products formed have the mass numbers 31 or 45, apparently a methylamine or a dimethylamine as neutral products are formed. These reactions correspond to the formation processes of the primary ions. The intense fragment ions $(CH_3)_2PCH_2^+$ and $(CH_3)_2P^+$ are $(M - CH_3NH^{\cdot})^+$ and $(M - (CH_3)_2N^{\cdot})^+$ ions. The molecular ion is unreactive as is the molecular ion of the trimethylphosphinimine, only collision dissociation reactions are observed. Contrary, the protonated molecule, N-methyl-trimethylaminophosphonium is very reactive. All the product ions observed in the spectrum are formed by reactions of this ion, they are summarized in Figure 7. As indicated in this figure the first step of the reactions is a nucleophilic addition of the neutral $(CH_3)_3PNCH_3$ molecule to the protonated molecule.

The product ions observed in the spectrum are formed by different dissociation processes of this intermediate. Partially, the same product ions are also formed by addition reactions of fragment ions to the neutral molecule. The ion m/e 180 is formed by dissociation of methylamine from the intermediate, shown in Figure 7:

$$(CH_3)_3PN(CH_3)H^+ + (CH_3)_3PNCH_3 \longrightarrow (CH_3)_3PNP(CH_3)_2CH_2^+ + CH_3NH_2 \quad (4.39)$$

and by an addition reaction of the fragment ion m/e 75:

$$(CH_3)_2PCH_2^+ + (CH_3)_3PNCH_3 \longrightarrow (CH_3)_3PNP(CH_3)_2CH_2^+ \quad (4.38)$$

Table 4. Ion-Molecule Reactions of N-methyl-trimethyliminophsophorane

m/e	Product Ion	m/e	Primary Ion	No. of Reaction
41	$C_3H_5^+$	105	$(CH_3)_3PNCH_3^{\cdot+}$	(4.1)
42	$(CH_2)_2N^+$	105	$(CH_3)_3PNCH_3^{\cdot+}$	(4.2)
44	$(CH_3)_2N^+$	105	$(CH_3)_3PNCH_3^{\cdot+}$	(4.3)
45	PCH_2^+	105	$(CH_3)_3PNCH_3^{\cdot+}$	(4.4)
47	$H_2PCH_2^+$	75	$(CH_3)_2PCH_2^+$	(4.5)
		105	$(CH_3)_3PNCH_3^{\cdot+}$	(4.6)
57	$(CH)_2P^+$	105	$(CH_3)_3PNCH_3^{\cdot+}$	(4.7)
59	$(CH_2)_2P^+$	105	$(CH_3)_3PNCH_3^{\cdot+}$	(4.8)
60	$PNCH_3^+$	105	$(CH_3)_3PNCH_3^{\cdot+}$	(4.9)
61	$(CH_3)_2P^+$	76	$(CH_3)_3P^{\cdot+}$	(4.10)
		105	$(CH_3)_3PNCH_3^{\cdot+}$	(4.11)
62	$H_2PNCH_3^+$	105	$(CH_3)_3PNCH_3^{\cdot+}$	(4.12)
71	$PC_3H_4^+$	106	$(CH_3)_3PN(CH_3)H^+$	(4.13)
73	$PC_3H_6^+$	105	$(CH_3)_3PNCH_3^{\cdot+}$	(4.14)
75	$(CH_3)_2PCH_2^+$	105	$(CH_3)_3PNCH_3^{\cdot+}$	(4.15)
76	$(CH_3)_3P^{\cdot+}$	106	$(CH_3)_3PN(CH_3)H^+$	(4.16)
77	$(CH_3)_3PH^+$	105	$(CH_3)_3PNCH_3^{\cdot+}$	(4.17)
90	$(CH_3)_2PNCH_3^+$	76	$(CH_3)_3P^{\cdot+}$	(4.18)
		106	$(CH_3)_3PN(CH_3)H^+$	(4.19)
104	$(CH_3)_3PNCH_2^+$	61	$(CH_3)_2P^+$	(4.20)
		75	$(CH_3)_2PCH_2^+$	(4.21)
105	$(CH_3)_3PNCH_3^{\cdot+}$	75	$(CH_3)_2PCH_2^+$	(4.22)
106	$(CH_3)_3PN(CH_3)H^+$	76	$(CH_3)_3P^{\cdot+}$	(4.23)
		105	$(CH_3)_3PNCH_3^{\cdot+}$	(4.24)
119	$(CH_2)_2PP(CH_2)_2H^+$	45	PCH_2^+	(4.25)
		47	$H_2PCH_2^+$	(4.26)
		61	$(CH_3)_2P^+$	(4.27)
		106	$(CH_3)_3PN(CH_3)H^+$	(4.28)
121	$(CH_2)_2PP(CH_3)_2H^+$	47	$H_2PCH_2^+$	(4.29)
		90	$(CH_3)_2PNCH_3^+$	(4.30)
		106	$(CH_3)_3PN(CH_3)H^+$	(4.31)
135	$(CH_2)_2PP(CH_3)_3^+$	61	$(CH_3)_2P^+$	(4.32)
		76	$(CH_3)_3P^{\cdot+}$	(4.33)

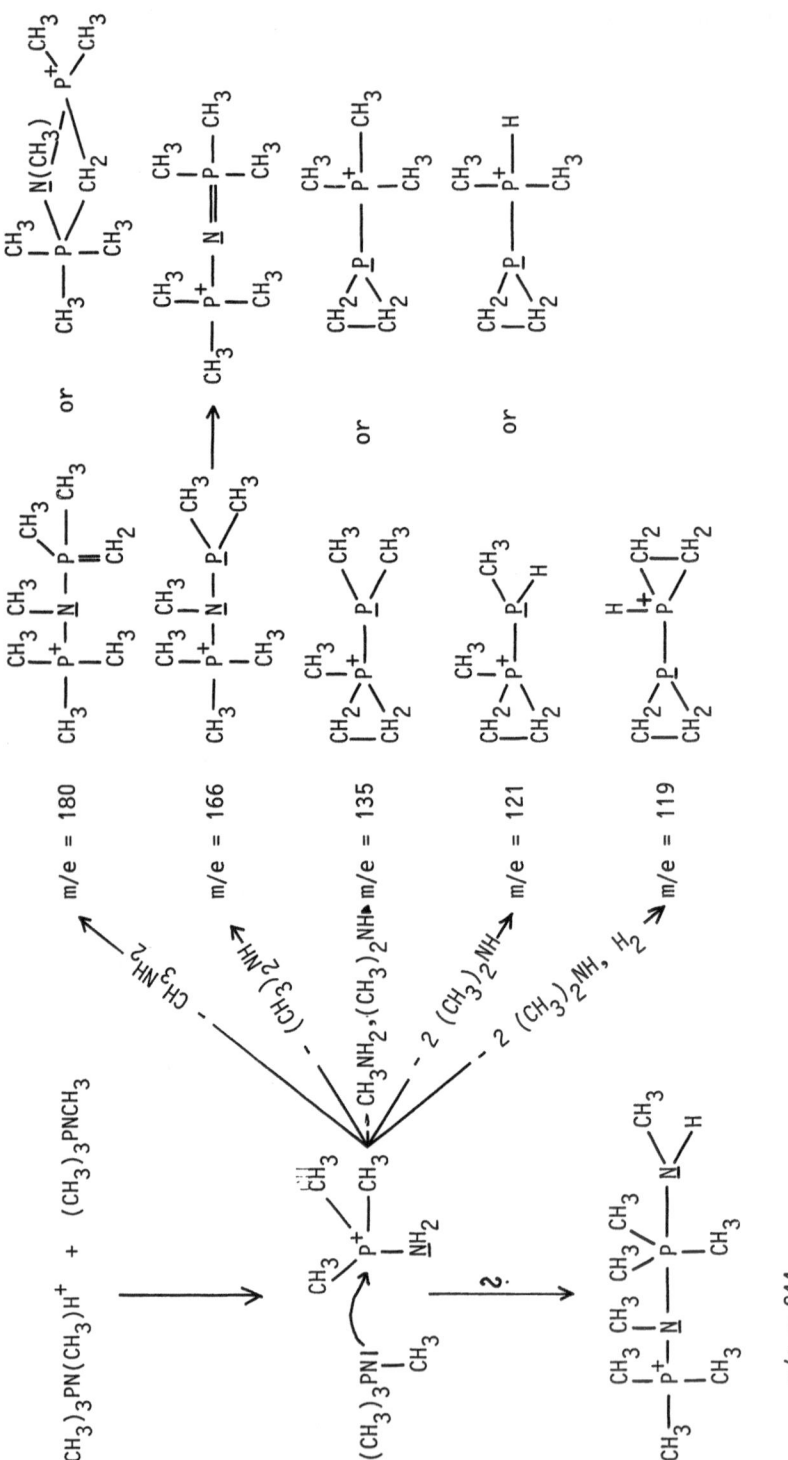

Fig. 7. Ion-molecule reactions of the protonated molecule of N-methyl-trimethyliminophosphorane

Table 4 (cont'd)

m/e	Product Ion	m/e	Primary Ion	No. of Reaction
135	$(CH_2)_2PP(CH_3)_3^+$	106	$(CH_3)_3PN(CH_3)H^+$	(4.34)
166	$(CH_3)_3PNP(CH_3)_3^+$	61	$(CH_3)_2P^+$	(4.35)
		76	$(CH_3)_3P^{\cdot+}$	(4.36)
		106	$(CH_3)_3PN(CH_3)H^+$	(4.37)
180	$(CH_3)_3PNP(CH_3)_2CH_2^+$	75	$(CH_3)_2PCH_2^+$	(4.38)
		106	$(CH_3)_3PN(CH_3)H^+$	(4.39)

In the course of reaction (4.39) the terminal amino group of the intermediate is most likely dissociated, because a hydrogen atom and a methyl group are already bonded to that nitrogen atom. The transfer of a proton from a methyl group to a nitrogen may easily be performed (transylidation reaction). The product ion m/e 180 has a stable phosphonium structure. The ion, however contains a phosphonium and an ylid center and a further nucleophilic attack should occur yielding an ion with a ring structure.

$$(CH_3)_2\overset{+}{P}\!-\!\bar{N}(CH_3)\!-\!P(CH_3)_3 \quad\longrightarrow\quad (CH_3)_2\underset{\|}{P}\!-\!\bar{N}(CH_3)\!-\!P(CH_3)_3$$

$$\longrightarrow \quad (CH_3)_2\underset{\|}{P}\!-\!\bar{N}(CH_3)\!-\!P^+(CH_3)_3 \quad + \quad CH_3NH_2$$

$$\longrightarrow \quad (CH_3)_2\overset{+}{P} \overset{\bar{N}(CH_3)}{\underset{CH_2}{\diagup\!\!\!\diagdown}} P(CH_3)_3 \quad + \quad CH_3NH_2$$

The product ion m/e 166 is formed by reactions of the primary ions m/e 61, 76 and 106:

$$(CH_3)_2P^+ \; + \; (CH_3)_3PNCH_3 \longrightarrow (CH_3)_3PNP(CH_3)_3^+ \qquad (4.35)$$

$$(CH_3)_3P^{\cdot+} \; + \; (CH_3)_3PNCH_3 \longrightarrow (CH_3)_3PNP(CH_3)_3^+ \; + \; CH_3^{\cdot} \quad (4.36)$$

$$(CH_3)_3PN(CH_3)H^+ \; + \; (CH_3)_3PNCH_3 \longrightarrow (CH_3)_3PNP(CH_3)_3^+ \; + \; (CH_3)_2NH \;(4.37)$$

The most probable structure of the product ion is the resonance-stabilized structure already discussed earlier. The ion is stable and known in the chemistry of the condensed phase. Salts of it were prepared by Schmidbaur and Jonas[15] and by Rave[17]. Its structure, bis-trimethyl-phosphoranylideneammonium, was proved by IR and NMR spectroscopy. Salts of this ion can be prepared by reaction of the lithium derivative of trimethylphosphinimine with dimethylchlorophosphine and methylation of the reaction product:

$$(CH_3)_3PNP(CH_3)_2 \; + \; CH_3J \longrightarrow ((CH_3)_3PNP(CH_3)_3)^+ \; J^-$$

Rave prepared salts of the ion by thermolysis of aminotrimethylphosphonium salts. This method of preparation indicates the great stability of the ion m/e 166.

In the reaction (4.37) of the even-electron ion $(CH_3)_3PNCH_3H^+$ also an even-electron neutral molecule, dimethylamine, is formed. In the case of the radical ion, m/e 76, a neutral radical, $CH_3^\cdot$, is dissociated (4.36)

For the product ion m/e 135 a structure without a nitrogen atom is assumed. In its formation reaction by the protonated molecule:

$$(CH_3)_3PN(CH_3)H^+ + (CH_3)_3PNCH_3 \longrightarrow (CH_2)_2PP(CH_3)_3^+ + CH_3NH_2 + (CH_3)_2NH$$
$$(4.34)$$

dimethylamine and methylamine are dissociated as neutral molecules. In the reaction (4.32)

$$(CH_3)_2P^+ \; + \; (CH_3)_3PNCH_3 \longrightarrow (CH_2)_2PP(CH_3)_3^+ \; + \; CH_3NH_2 \qquad (4.32)$$

a methylamine molecule is dissociated. The ion m/e 76 reacts to the product ion m/e 135, too, indicating similar structures of the ions m/e 61 and 76, $(CH_3)_2P^+$ and $(CH_3)_3P^{\cdot +}$.

$$(CH_3)_3P^{\cdot +} \; + \; (CH_3)_3PNCH_3 \longrightarrow (CH_2)_2PP(CH_3)_3^+ \; + \; CH_3^\cdot \; + \; CH_3NH_2 \qquad (4.33)$$

Contrary to the formation of the ion m/e 166 from the protonated molecule, reaction (4.34) includes rearrangement, now the nitrogen in the bridge is engaged in the formation of the neutral methylamine. Furthermore, two hydrogen atoms are dissociated from the two methyl groups at the phosphorus atom and the formation of a three-membered ring containing this phosphorus atom may be assumed. Similar H_2 dissociations have already been observed in the formation of primary ions and in the ion chemistry of trimethylphosphine[13].

Aside from the ion m/e 137, the most intense product ion in the ion chemistry of trimethylphosphine, the ion m/e 135 is formed by an addition reaction of $(CH_2)_2P^+$ to the neutral $(CH_3)_3P$. The product ion has the structure:

$$(CH_3)_3P^+ \!\!-\!\! \bar{P} \!\!<^{\displaystyle CH_2}_{\displaystyle CH_2}$$

The product ions of the reactions (4.32) - (4.34) may have the same structure or the isomeric structure:

$$(CH_3)_3P^+ \!\!-\!\! \bar{P} \!\!<^{\displaystyle CH_2}_{\displaystyle CH_2} \qquad\qquad (CH_3)_2\bar{P} \!\!-\!\! \overset{\displaystyle CH_3}{\underset{}{P^+}} \!\!<^{\displaystyle CH_2}_{\displaystyle CH_2}$$

Replacement of a methyl group in the ion m/e 61 by a hydrogen atom leads to the ion m/e 47, $HPCH_3^+$. This ion undergoes a reaction corresponding to reaction (4.32) of $(CH_3)_2P^+$:

$$HPCH_3^+ \;+\; (CH_3)_3PNCH_3 \longrightarrow (CH_2)_2PP(CH_3)_2H^+ \;+\; CH_3NH_2 \qquad (4.29)$$

Again a neutral methylamine molecule is dissociated and one methyl group in the product ion m/e 135 is replaced by a hydrogen atom as it is in the reacting ion. This leads to the conclusion that the two hydrogen atoms, dissociated in the neutral methylamine come from the neutral reactant. Therefore the structure:

$$\begin{array}{c} H \\ H_3C \end{array}\!\!\!\!>\bar{P}\!-\!\overset{+}{P}\!<\!\!\!\begin{array}{c} CH_2 \\ | \quad | \\ CH_2 \end{array}$$

is more probable.

The product ion m/e 121 is formed by two further ion-molecule reactions

$$(CH_3)_2PNCH_3^+ + (CH_3)_3PNCH_3 \longrightarrow (CH_2)_2PP(CH_3)_2H^+ + CH_3NH_2 + C_2H_4NH \quad (4.30)$$

$$(CH_3)_3PN(CH_3)H^+ + (CH_3)_3PNCH_3 \longrightarrow (CH_2)_2PP(CH_3)_2H^+ + 2 (CH_3)_2NH \quad (4.31)$$

In the reaction (4.30) a neutral molecule with the composition C_2H_5N is
dissociated. The most probable structure of this molecule is $(CH_2)_2NH$.

The structure of the product ion m/e 119 is assumed to be similar
to that of the product ion m/e 121, because the formation reactions are
similar, too.

$$PCH_2^+ \qquad\qquad + (CH_3)_3PNCH_3 \longrightarrow (CH_2)_2PP(CH_2)_2H^+ + CH_3NH_2 \qquad (4.25)$$

$$HPCH_3^+ \qquad\qquad + (CH_3)_3PNCH_3 \longrightarrow (CH_2)_2PP(CH_2)_2H^+ + CH_3NH_2 + H_2 (4.26)$$

$$(CH_3)_2P^+ \qquad\quad + (CH_3)_3PNCH_3 \longrightarrow (CH_2)_2PP(CH_2)_2H^+ + (CH_3)_2NH + H_2 (4.27)$$

$$(CH_3)_3PN(CH_3)H^+ + (CH_3)_3PNCH_3 \longrightarrow (CH_2)_2PP(CH_2)_2H^+ + 2(CH_3)_2NH + 2H_2 (4.28)$$

Trimethylphosphine oxide

The ion-molecule reactions of trimethylphosphine oxide are listed in
Table 5. The chemistry of the neutral molecule shows a large basicity
of the oxygen atom site. Therefore a nucleophilic attack of the oxygen
atom of the neutral phosphine oxide at the position of the charge of the
ion is assumed as first step in the ion-molecule reactions.

The molecular ion undergoes only two ion-molecule reactions. The
protonated molecule is again very reactive. It has again a phosphonium
ion structure $(CH_3)_3POH^+$. The differences in reactivity between the protonated molecule and   the molecular ion may be rationalized by a de-

Table 5. Ion-Molecule Reactions of Trimethylphosphine Oxide

m/e	Product Ion	m/e	Primary Ion	No. of Reaction
43	PC^+	77	$(CH_3)_2PO^+$	(5.1)
44	$PCH^{\cdot+}$	77	$(CH_3)_2PO^+$	(5.2)
45	PCH_2^+	47	PO^+	(5.3)
		77	$(CH_3)_2PO^+$	(5.4)
		92	$(CH_3)_3PO^{\cdot+}$	(5.5)
46	$PCH_3^{\cdot+}$	77	$(CH_3)_2PO^+$	(5.6)
		92	$(CH_3)_3PO^{\cdot+}$	(5.7)
47	PO^+	49	H_2PO^+	(5.8)
		75	$(CH_2)_2PO^+$	(5.9)
		77	$(CH_3)_2PO^+$	(5.10)
48	$POH^{\cdot+}$	77	$(CH_3)_2PO^+$	(5.11)
49	H_2PO^+	47	PO^+	(5.12)
		77	$(CH_3)_2PO^+$	(5.13)
		92	$(CH_3)_3PO^{\cdot+}$	(5.14)
57	$P(CH)_2^+$	47	PO^+	(5.15)
		49	H_2PO^+	(5.16)
		77	$(CH_3)_2PO^+$	(5.17)
		92	$(CH_3)_3PO^{\cdot+}$	(5.18)
58	$P(CH)_2H^{\cdot+}$	77	$(CH_3)_2PO^+$	(5.19)
		92	$(CH_3)_3PO^{\cdot+}$	(5.20)
59	$P(CH_2)_2^+$	47	PO^+	(5.21)
		49	H_2PO^+	(5.22)
		77	$(CH_3)_2PO^+$	(5.23)
		92	$(CH_3)_3PO^{\cdot+}$	(5.24)
61	CH_2PO^+	47	PO^+	(5.25)
		49	H_2PO^+	(5.26)
		77	$(CH_3)_2PO^+$	(5.27)
		92	$(CH_3)_3PO^{\cdot+}$	(5.28)
62	$CH_3PO^{\cdot+}$	49	H_2PO^+	(5.29)
		77	$(CH_3)_2PO^+$	(5.30)
		92	$(CH_3)_3PO^{\cdot+}$	(5.31)
63	$(CH_3)HPO^+$	47	PO^+	(5.32)
		49	H_2PO^+	(5.33)

Table 5 (cont'd)

m/e	Product Ion	m/e	Primary Ion	No. of Reaction
63	$(CH_3)HPO^+$	77	$(CH_3)_2PO^+$	(5.34)
		92	$(CH_3)_3PO^{\cdot+}$	(5.35)
75	$(CH_2)_2PO^+$	49	H_2PO^+	(5.36)
		93	$(CH_3)_3POH^+$	(5.37)
77	$(CH_3)_2PO^+$	47	PO^+	(5.38)
		49	H_2PO^+	(5.39)
		93	$(CH_3)_3POH^+$	(5.40)
92	$(CH_3)_3PO^{\cdot+}$	77	$(CH_3)_2PO^+$	(5.41)
93	$(CH_3)_3POH^+$	49	H_2PO^+	(5.42)
		77	$(CH_3)_2PO^+$	(5.43)
		92	$(CH_3)_3PO^{\cdot+}$	(5.44)
107	$(CH_3)_3POCH_3^+$	77	$(CH_3)_2PO^+$	(5.45)
		93	$(CH_3)_3POH^+$	(5.46)
123	$(CH_3)_3POP^+$	59	$P(CH_2)_2^+$	(5.47)
		77	$(CH_3)_2PO^+$	(5.48)
		93	$(CH_3)_3POH^+$	(5.49)
139	$(CH_3)_3POPO^+$	47	PO^+	(5.50)
		62	$CH_3PO^{\cdot+}$	(5.51)
		75	$(CH_2)_2PO^+$	(5.52)
		77	$(CH_3)_2PO^+$	(5.53)
		93	$(CH_3)_3POH^+$	(5.54)
149	$(CH_3)_3POP(CH)_2^+$	57	$P(CH)_2^+$	(5.55)
		77	$(CH_3)_2PO^+$	(5.56)
		93	$(CH_3)_3POH^+$	(5.57)
151	$(CH_3)_2POP(CH_2)_2^+$	57	$P(CH)_2^+$	(5.58)
		59	$P(CH_2)_2^+$	(5.59)
		77	$(CH_3)_2PO^+$	(5.60)
		93	$(CH_3)_3POH^+$	(5.61)
153	$(CH_3)_3POP(CH_3)_2^+$	77	$(CH_3)_2PO^+$	(5.62)
		93	$(CH_3)_3POH^+$	(5.63)
167	$(CH_3)_3POP(CH_2)_2O^+$	75	$(CH_2)_2PO^+$	(5.64)
		77	$(CH_3)_2PO^+$	(5.65)
169	$(CH_3)_3POP(CH_3)_2O^+$	77	$(CH_3)_2PO^+$	(5.66)
		92	$(CH_3)_3PO^{\cdot+}$	(5.67)
185	$(CH_3)_3POP(CH_3)_3OH^+$	93	$(CH_3)_3POH^+$	(5.68)

crease of the phosphonium character of the latter ion.

$$(CH_3)_3P^+ - \underline{\underline{O}}^{\boldsymbol\cdot} \longleftrightarrow (CH_3)_3P = \underline{\underline{O}}^{\boldsymbol\cdot +}$$

The molecular ion reacts to the protonated molecule and to the product ion m/e 169:

$$(CH_3)_3PO^{\boldsymbol\cdot +} + (CH_3)_3PO \longrightarrow (CH_3)_3POP(CH_3)_2O^+ + CH_3^{\boldsymbol\cdot} \qquad (5.67)$$

The same product ion is formed by an addition reaction of the ion m/e 77, $(CH_3)_2PO^+$:

$$(CH_3)_2PO^+ + (CH_3)_3PO \longrightarrow (CH_3)_3POP(CH_3)_2O^+ \qquad (5.65)$$

The shown structure of the ion is therefore the most probable.

The ion m/e 77 is the most abundant ion in the ICR mass spectrum and the most reactive in its ion chemistry. Its reactions are shown in Figure 8. The first step in the reactions is a nucleophilic attack of the neutral trimethylphosphine oxide at the phosphonium site of the ion, leading to a P-O-P bridged intermediate m/e 169, $(CH_3)_3POP(CH_3)_2O^+$, which stabilizes (reaction (5.66)) or dissociates to form products with lower molecular weights. It could not excluded that rearrangements occur during this dissociation to form phosphonium substituted phosphine oxides. It is known from the ion chemistry of the phosphines[13,18] that they attack with their lone pair electrons phosphonium ions. Such a nucleophilic attack would yield a cyclic intermediate in the reactions of the ion m/e 77:

$$(CH_3)_3P^+ \overset{\displaystyle\frown}{\underset{\underline{\underline{O}}}{}} P(CH_3)_2 \longrightarrow (CH_3)_3P \overset{}{\underset{\underline{\underline{O}}}{\diagdown\diagup}} P^+(CH_3)_2 \longrightarrow (CH_3)_3P^+ - \underset{\underset{\displaystyle \underline{\underline{O}}}{\|}}{P}(CH_3)_2$$

Analogous to the Michaelis-Arbuzov rearrangement a tertiary phosphine oxide is formed, substituted with a phosphonium ion. All the product ions which can be expected by formal addition of primary ions to the

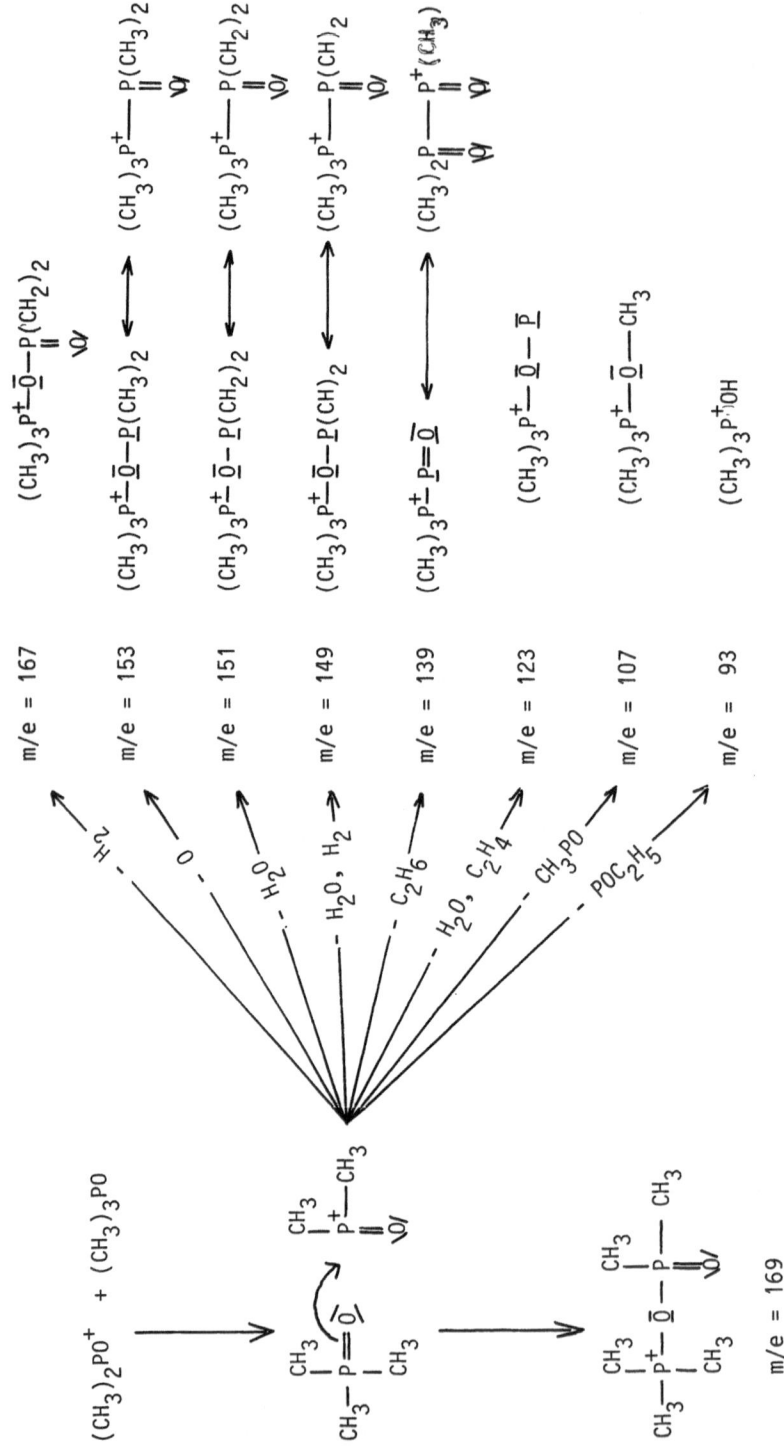

Fig. 8. Ion-molecule reactions of the fragment ion m/e 77, $(CH_3)_2PO^+$, of trimethylphosphine oxide

neutral phosphine oxide are detected in the ion chemistry of the ion m/e 77 . In the course of reaction (5.45) $(CH_3)_2PO^+$ transfers a methyl cation to the neutral phosphine oxide.

$$(CH_3)_2PO^+ + (CH_3)_3PO \longrightarrow (CH_3)_3POCH_3^+ + CH_2PO \qquad (5.45)$$

The ion m/e 107 is the single product ion formed by an alkylation reaction at the oxygen atom. It is known from the chemistry of the phosphine oxides that alkylation at the oxygen atom is an anusual reaction[19].

The structure of the fragment ion m/e 75 is similar to that of the ion m/e 77. It forms several of the product ions already observed at the ion m/e 77

$$(CH_2)_2PO^+ + (CH_3)_3PO \longrightarrow (CH_3)_3POP(CH_2)_2O^+ \qquad (5.64)$$

The product ion m/e 139 is formed by a transfer of a PO^+ ion by the $(CH_2)_2PO^+$ ion. This reaction may indicate a cyclic structure of the ion m/e 75 as has already been discussed for the ion m/e 75 in the ion chemistry of trimethylmethylenephosphorane (2.34):

$$(CH_2)_2PO^+ + (CH_3)_3PO \longrightarrow (CH_3)_3POPO^+ + C_2H_4 \qquad (5.52)$$

This product ion is also formed by a reaction of the fragment ion m/e 62

$$CH_3PO^{\cdot +} + (CH_3)_3PO \longrightarrow (CH_3)_3POPO^+ + CH_3^{\cdot} \qquad (5.51)$$

in this case a methyl radical is dissociated.

The fragment ion m/e 59, $(CH_2)_2P^+$, shows an addition reaction:

$$(CH_2)_2P^+ + (CH_3)_3PO \longrightarrow (CH_3)_3POP(CH_2)_2^+ \qquad (5.59)$$

and the reaction which is characteristic for its cyclic structure, transfer of a P^+ ion and dissociation of a neutral C_2H_4 molecule:

$$(CH_2)_2P^+ + (CH_3)_3PO \longrightarrow (CH_3)_3POP^+ + C_2H_4 \qquad (5.47)$$

An analogous addition reaction is observed for the ion m/e 57, $(CH)_2P^+$:

$$(CH)_2P^+ + (CH_3)_3PO \longrightarrow (CH_3)_3POP(CH)_2^+ \qquad (5.55)$$

A further addition reaction is observed for the ion m/e 47

$$PO^+ \quad + \quad (CH_3)_3PO \quad \longrightarrow \quad (CH_3)_3POPO^+ \qquad (5.50)$$

The protonated molecule is formed by reactions of the ions m/e 49, 77 and 92

$$H_2PO^+ \quad + \quad (CH_3)_3PO \quad \longrightarrow \quad (CH_3)_3POH^+ \; + \; HPO \qquad (5.42)$$

$$(CH_3)_2PO^+ \; + \; (CH_3)_3PO \quad \longrightarrow \quad (CH_3)_3POH^+ \; + \; POC_2H_5 \qquad (5.43)$$

$$(CH_3)_3PO^{\cdot+} \; + \; (CH_3)_3PO \quad \longrightarrow \quad (CH_3)_3POH^+ \; + \; POC_3H_8^{\cdot} \qquad (5.44)$$

The protonated molecule is the most intense product ion, it is a stable phosphonium ion. Its ion-molecule reactions are shown in Figure 9. The intermediate formed by a nucleophilic attack of the neutral phosphine oxide at the site of the charge of the ion can either stabilize (reaction (5.68)) or dissociate and form the reaction products shown in the figure. In the first step of the reaction a P-O-P bridge is formed. Contrary the intermediate in the reaction of the molecular ion is very unstable and the dimer molecule could not be detected in the spectrometer. This intermediate stabilizes by dissociation of a methyl radical, the ion m/e 169 is formed:

$$(CH_3)_3PO^{\cdot+} \; + \; (CH_3)_3PO \quad \longrightarrow \quad (CH_3)_3POP(CH_3)_2O^+ \; + \; CH_3^{\cdot} \qquad (5.67)$$

This dissociation is not detected for the intermediate in the reaction of the protonated molecule, which is an even-electron ion and should not dissociate a radical. However, dissociation of methanol is observed, the product ion m/e 153 is formed:

$$(CH_3)_3POH^+ \; + \; (CH_3)_3PO \quad \longrightarrow \quad (CH_3)_3POP(CH_3)_2^+ \; + \; CH_3OH \qquad (5.63)$$

The product ion m/e 153 may rearrange to a phosphine oxide in a Michaelis-Arbuzov-like rearrangement. The mechanism of the reaction and the dissociation of CH_3OH is in accord with the following reaction detected

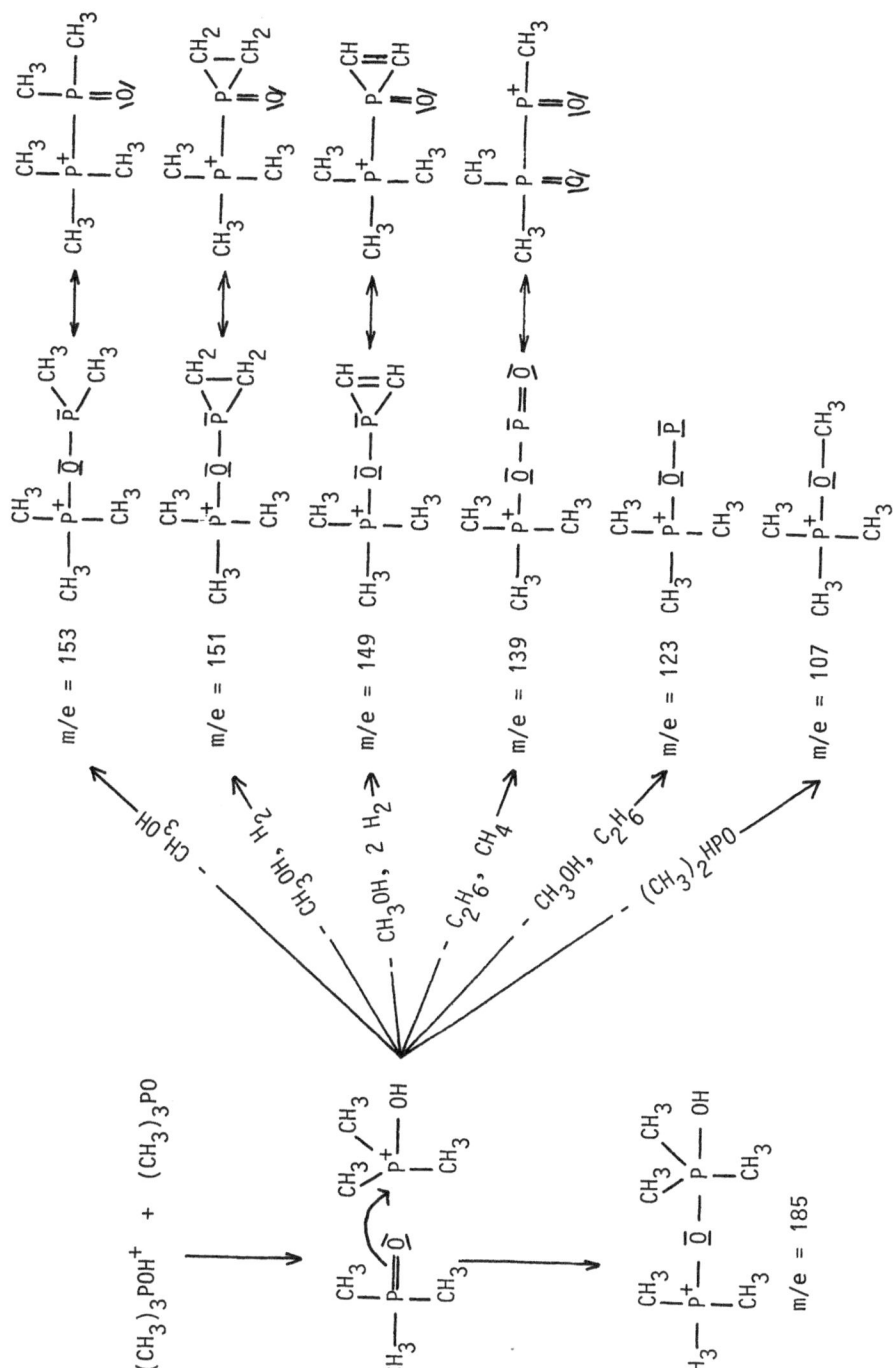

Fig. 9. Ion-molecule reactions of the protonated molecule of trimethylphosphine oxide.

in a mixture of trimethylphosphine oxide and tetramethyldiphosphine:

$$(CH_3)_2PP(CH_3)_2^{\cdot +} + (CH_3)_3PO \longrightarrow (CH_3)_3POP(CH_3)_2^+ + (CH_3)_2P^{\cdot}$$

In the reactions (5.57) and (5.61) the product ions m/e 149 and 151 are formed. They contain H_2 and 2 H_2, respectively, less than the product ion m/e 153:

$$(CH_3)_3POH^+ + (CH_3)_3PO \rightarrow (CH_3)_3POP(CH_2)_2^+ + CH_3OH + H_2 \qquad (5.57)$$

$$(CH_3)_3POH^+ + (CH_3)_3PO \rightarrow (CH_3)_3POP(CH)_2^+ + CH_3OH + 2\ H_2 \qquad (5.61)$$

Sequences of ions differing in two mass units are well known in the formation of primary and product ions of phosphines and phosphoranes with two methyl groups.

Further product ions are m/e 107, 123 and 139:

$$(CH_3)_3POH^+ + (CH_3)_3PO \longrightarrow (CH_3)_3POCH_3^+ + (CH_3)_2HPO \qquad (5.46)$$

$$(CH_3)_3POH^+ + (CH_3)_3PO \longrightarrow (CH_3)_3POP^+ + CH_3OH + C_2H_6 \qquad (5.49)$$

$$(CH_3)_3POH^+ + (CH_3)_3PO \longrightarrow (CH_3)_3POPO^+ + C_2H_6 + CH_4 \qquad (5.54)$$

Aside from the reactions discussed there are further ion-molecule reactions in the ion chemistry of trimethylphosphine oxide, which can be interpreted as collision dissociations. They are also listed in Table 5. The ion m/e 75, $(CH_2)_2PO^+$, is only formed by such reactions. It is not detected in the primary ion spectrum.

Discussion

The ylids $(CH_3)_3PCH_2$, $(CH_3)_3PNH$, $(CH_3)_3PO$ are isoelectronic compounds with increasing electronegativity of the basic center in the order $(CH_3)_3PCH_2$, $(CH_3)_3PNH$, $(CH_3)_3PO$. $(CH_3)_3PNCH_3$ is more stable than its N-methyl derivative, because only substituents with negative induction effects can delocalize the negative charge at the ylidic nitrogen atom and thus stabilize the ylid.

With the exception of trimethylphosphine oxide in the mass spectra of all the ylids the molecular ions are the most abundant ions. Most of the fragment ions contain again the phosphorus atom, which is normally the site of the charge. The most intense fragment ions are those formed by dissociation of a radical from the molecular ion. All the molecular ions dissociate a methyl radical, forming the intense fragment ions $(CH_3)_2PCH_2^+$, $(CH_3)_2PNH^+$, $(CH_3)_2PNCH_3^+$ and $(CH_3)_2PO^+$. Only in the case of N-methyl-trimethyliminophosphorane CH_3NH dissociation is favoured. Apparently, stabilization of the CH_3NH radical by the methyl group is important because a corresponding dissociation of a NH_2 radical is almost completely missed at the trimethylphosphine imine.

All the phsophonium ions with two methyl groups can dissociate a neutral ethene molecule. The methyl groups are replaced by hydrogen atoms, the structure of the molecule is probably preserved.

Groups of ions differing by two or four mass units are abundant in the mass spectra and the ion-molecule reactions. In the case of the $(CH_2)_2PNH^+$ the ion $(CH)_2PNH^+$ is not detected, it is assumed that NH_3 is dissociated and the ion ion $(CH)_2P^+$ is formed.

The ion $C_3H_5^+$ is observed in the mass spectra of $(CH_3)_3PCH_2$, $(CH_3)_3PNH$ and $(CH_3)_3P^6$. The appearance of this ion shows the similarity of the decomposition processes of the three compounds.

Ylids are very polar and reactive compounds. Therefore the first step in the ion-molecule reactions is a nucleophilic attack of the negatively polarized site of the neutral molecule at the positive center of the ion. Contrary, only very few reactions occur via attack at a carbon atom of the ion.

The intensity ratios of the molecular and fragment ions and the protonated molecules are normally reproduced in the ratios of the numbers of ion-molecule reactions underwent by the respective ions.

The product ions can be divided into ions which are formally formed by addition of fragment ions to the neutral molecules and ions not formed by such reactions. The first-named are the more abundant. All the product ions with high m/e ratios belong to the first group. These ions are in part formed by addition reactions of the respective fragment ions and in part by reactions of the molecular ions or the protonated molecules. In the course of these reactions neutral particles are formed which are similar to the fragments formed during ionization by electron impact and subsequent fragmentation. Product ions with low m/e ratios belong to the second class. In the course of the formation reactions of these ions neutral products are formed which normally differ from the fragments known from the mass spectra. Therefore the structures of the ions of the second class should be different from those of the first class.

As can be expected, the protonated molecules are the most abundant or among the most abundant product ions. The protonated molecules are mainly formed by reactions of the molecular ions. With increasing electronegativity of the negative center of the ylids the abundance of the dimer molecule ions or the protonated dimer molecules increases. In the case of trimethylphosphine oxide the protonated dimer is the most abundant product ion.

The product ions of trimethylmethylenephosphorane can be subsumed under the general formula $(CH_3)_3PCHPR_n^+$ ($R = CH_3$, H; $n = 1,2,3$). The most

important of these is the resonance-stabilized ion $(CH_3)_3PCHP(CH_3)_3^+$. A
second important class of products are the ions formed by methylenetrans-
fer reactions, a methylene group is transferred by the neutral molecule
to the molecular or fragment ion, neutral $(CH_3)_3P$ is formed. This type
of reaction is only observed for $(CH_3)_3PCH_2$.

Important reactions of the first kind in the ion chemistry of
$(CH_3)_3PNH$ are addition reactions of the fragment ion $(CH_3)_2PNH^+$ and the
protonated molecule to the neutral molecule. For the second class of reactions again rearrangements including formation of P-P and dissociation
of P-N bonds must be assumed. The product ion m/e 137 contains only P,
C and H atoms. In the case of the heavy product ion m/e 167, the nitrogen remains in the product ion, because it is a part of the resonance
system.

$$(CH_3)_3P^+ \!\!-\! \bar{N}\!=\!P(CH_3)_3 \quad \longleftrightarrow \quad (CH_3)_3P\!=\!\bar{N}\!-\!P^+(CH_3)_3$$

This behaviour is even more pronounced in the ion-molecule reactions
of N-methyl-trimethyliminophosphorane. In the mass spectrum the ion m/e
75 is the most abundant fragment ion. The nitrogen atom has been dissociated in the neutral CH_3NH radical. Formation of nitrogen containig
fragment ions is only of low probability, as indicated by the low intensity of the ion m/e 90.

Numerous ion-molecule reactions result in the formation of neutral
methylamine molecules. The basicity of the nitrogen atom in the N-methyl-
trimethyliminophosphorane is increased compared with the NH derivative.
The increased basicity may be the reason that in the reactions of the
N-methyl-trimethyliminophosphorane methylamine and in the reactions of
the NH derivative only NH is dissociated. Aside from methylamine also

dimethylamine is formed in the first case. As already mentioned for $(CH_3)_3PNH$, product ions with m/e ratios lower than 135 contain newly formed P-P bonds, nitrogen containing neutral particles are dissociated. The product ions exclusively contain P, C and H.

The only three product ions which contain nitrogen are the protonated molecule, the ion m/e 166, which has the nitrogen atom included in the resonance system and the ion m/e 180. This ion is the second most intense product ion. It has probably a cyclic structure:

$$(CH_3)_3P \underset{CH_2}{\overset{\underset{N}{CH_3}}{\diagdown\diagup}} P^+(CH_3)_2$$

The product ions found in the ion-molecule reactions of trimethylphosphine oxide can also be divided into the two groups already mentioned: The first group are the ions formed by formal addition of fragment ions to the neutral molecule, the second are the remaining ions. In the second group there are the protonated molecule and the product ions m/e 107 and 123. With the exception of product ions from $(CH_3)_3PCH_2$ most of the product ions have even electron numbers. This is also true for phosphine oxide. The molecular ion $(CH_3)_3PO^{\cdot +}$ undergoes only two ion-molecule reactions which yield product ions with masses greater than that of the molecular ion, the protonated molecule and the product ion m/e 169. In the latter the intermediate is stabilized by dissociation of a methyl radical.

The protonated molecule forms the majority of the product ions. In its reactions neutral products with even electron numbers are formed, like CH_3OH, carbon hydrides or H_2.

The reactivity of the most intense fragment ion m/e 77 is comparable to that of the protonated molecule. Now in the course of the reactions a water molecule is dissociated instead of a methanol molecule. This leads to the conclusion that the neutral reaction products are chiefly dissociated from the reacting ion.

All the product ions contain a structure $(CH_3)_3P^+OP$. The only exceptions are the protonated molecule and the ion m/e 107 which is formed by alkylation at the oxygen.

In the ions $(CH_3)_3P^+OPR'$ the first phosphorus atom is a phosphonium center and the second can be either ter- or fivevalent, R'can be OR'', a hydroxy or an alkyl group. The product ions m/e 167 and 169 have the structures of tertiary phosphine oxides one substituent containing a phosphonium ion. For several further product ions a rearrangement to a phosphine oxide must be discussed. Two P-O single bonds are replaced by a P=O double bond and a P-P single bond, for example in the ion m/e 151

$$(CH_3)_3P^+ - \bar{\underline{O}} - \underline{P}(CH_2)_2 \qquad\qquad (CH_3)_3P^+ - P(CH_2)_2O$$

The bond energy of a P-O single bond is 92 kcal/mole[20], a P=O double bond has a bond energy of 139 kcal/mole[21] and a P-P single bond of 48 kcal/mole[22]. These bond energies do not exclude such a rearrangement. However, it should be mentioned that the bond energies were determined for neutral molecules.

Acknowledgements

One of us (H. H.) gratefully acknowledges the research facilities made available to him by the Deutsche Forschungsgemeinschaft. O.-R. H. thanks the Akademie der Wissenschaften und der Literatur, Mainz, for financial support.

References

1. W. Sawodny, Z. anorg. allg. Chem. <u>368</u>, 284 (1969)

2. H. Schmidbaur and W. Tronich, Chem. Ber. <u>101</u>, 604 (1968)

3. H. J. Bestmann and R. Zimmermann in Organic Phosphorus Compounds, vol. 3, chapt. 5A, Eds.: G. M. Kosolapoff and L. Maier, Wiley Interscience, London 1972

4. H. K. Wang, Acta Chem. Scand. <u>19</u>, 879 (1965)

5. O.-R. Hartmann, K.-P. Wanczek and H. Hartmann, Z. Naturforsch. <u>31a</u>, 630 (1976)

6. R. G. Gillis and G. L. Long, Org. Mass Spectrom. <u>2</u>, 1315 (1969)

7. O.-R. Hartmann, K.-P. Wanczek and H. Hartmann, unpublished results

8. K. A. Ostoja Starzewski, H. Tom Dieck and H. Bock, J. Organometal. Chem. <u>65,</u> 311 (1974)

9. J. Absar and R. J. van Wazer, J. Am. Chem. Soc. <u>94</u>,2382 (1972)

10. R. Hoffmann, D. B. Boyd and S. Z. Goldberg, J. Am. Chem. Soc. <u>92</u>, 3929 (1970)

11. H. Schmidbaur and W. Tronich, Chem. Ber. <u>101</u>, 3545 (1968)

12. Z.-C. Profous, K.-P. Wanczek, and H. Hartmann, Z. Naturforsch. <u>30a</u>, 1470 (1975)

13. K.-P. Wanczek and Z.-C. Profous, Intern. J. Mass Spectrom. Ion Phys. <u>17</u>,23 (1975)

14. D. F. Clemens, M. L. Caspar, D. Rosenthal and R. Peluso, Inorg. Chem. <u>9</u>, 960 (1970)

15. H. Schmidbaur and J. Jonas, Chem. Ber. <u>101</u>, 1271 (1968)

16. K. Issleib and W. Seidel, Chem. Ber. <u>92</u>, 2681 (1959)

17. T. W. Rave, J. Org. Chem. <u>32</u>, 3461 (1975)

18. K.-P. Wanczek, Z. Naturforsch. <u>30a</u>, 329 (1975)

19. D. B. Denney, D. Z. Denney and L. A. Wilson, Tetrahedron Letters 85 (1968)

20. S. B. Hartly, W. S. Holmes, J. K. Jackques, M. F. Mole and J. C. McCourbrey, Quart. Rev. (London) <u>43</u>, 204 (1965)

21. A. P. Claydon, P. A. Fowell and C. T. Mortimer, J. Chem. Soc. 3284 (1960)

22. A. F. Wells, Structural Inorganic Chemistry, 3rd edition, p. 236, Oxford Clarendon Press, Oxford 1962